PROPER FOOD COMBINING WORKS

Living Testimony

PROPER FOOD COMBINING W·O·R·K·S

Living Testimony

by Lee DuBelle

Warning–Disclaimer

This book is designed to provide information in regard to the subject matter covered. It is sold with the understanding that the publisher and author are not engaged in rendering medical or other professional services. If medical or professional assistance is required, those services should be sought with qualified persons according to your needs.

It is not the purpose of this book to reprint all the information that is available to the author and/or publisher, but to complement and supplement other texts. For more information see the order form on the last page.

The purpose of this book is to educate. The author and publisher shall have neither liability nor responsibility to any persons or entity with respect to any loss or damage caused, or alleged to be caused, directly or indirectly by the information contained in this book.

Revised and updated: 1987, 1988, 1989, 1990, 1992, 1993

7th Printing, 1993

Lee DuBelle, P.O. Box 35860, Phoenix, AZ 85069
(602) 863-2715

ISBN: 0-9618703-1-1

Cover and Page Design
by Seven Oaks Studio, Corrales, New Mexico
Typography and Book Production
by Walsh and Associates, Tempe, Arizona

Printed in the United States of America

Dedication

This book is dedicated to
my mother, Margaret.

Acknowledgement

Special thanks to
Roberta, Ken and Rosemary,

Table of Contents

Illustrations

I hope you ate your favorite meal today.
It's the last poorly combined meal you will ever
eat without feeling guilty.

INTRODUCTION

Over the past 25 to 30 years, Lee DuBelle has researched the nutritional ideas outlined in this book, using the most responsive testing equipment available: her own body and its health and well being.

The result is a system of combining foods that produces dynamic health, effective digestion and a high level of energy. It eliminates toxins in the body that can result in a wide range of maladies and illnesses, and reverses the harm we have all done to our bodies over years of eating improperly.

To see Lee DuBelle today, one would never believe that at one time in her life she suffered from obesity, mental and physical exhaustion, tuberculosis spots on her lungs, extremely low blood pressure, pre-menstrual syndrome, cysts on one ovary, bleeding stomach ulcers and a prolapsed colon. She also smoked three packs of cigarettes a day. Using the food combining techniques she discovered herself, she has stopped smoking and overcome all of her physical ailments. In addition, she has tremendous energy and travels the nation giving lectures and seminars on Proper Food Combining.

Proper Food Combining is the result of years of continuing personal research, consultations with clients on nearly every physical difficulty imaginable, and tapes of her seminars. She also has developed "Lee's Trim 1 & 2," which achieves better health through helping the body cleanse itself with herbs, rather than drugging it with medication.

Written in an easy-to-understand, conversational style, *Proper Food Combining* reviews every step of the body's digestive process with a minimum of scientific jargon, and indicates at each phase how the standard American diet not only contributes to, but produces, the low level of health generally experienced in this country.

Ken Bacher

C·H·A·P·T·E·R·1

THE ROAD TO FOOD COMBINING

Sometimes when I'm talking about my past health problems, it seems as though I'm talking about someone else. I often relate other people's experiences and I marvel at their recovery. Then I stop to realize that they weren't as ill as I was.

Almost 30 years ago, I was so exhausted mentally and physically that, even though I laughed a lot and put up a fairly good front, I was using every ounce of energy, patience, and good attitude that I could possibly pull together.

I was very poor, held a secretarial job, and just lived from payday to payday. I belonged to a health insurance program, through my job, that provided complete examinations for $1, plus the cost of medication. It seemed like every time I had a dollar, it went for one of those examinations. I used my sick leave as soon as it accumulated, and most often my vacation time also was used as sick leave.

At first, the examinations only showed things like, "perhaps an ulcer forming from stress." I was told to drink a lot of milk and cream. Constant backaches were determined to be "probably from stress." Constant legaches were determined to be "from stress." Even my obesity was determined to be "from stress."

We have all come to realize that stress is a large factor in a person's well being. But 30 years ago, it seemed to be used to suggest that one needed psychiatric care. One doctor even suggested that I start having an immoral affair every now and then to relieve the stress. I never followed his advice.

It's true that a lot of depression can come from being ill all of the time, but if people are trying to convince you that it is all in your head, find another source of help. We are finding more and more that even if it *is* a mental problem, it could be caused by physical deficiencies, or physical breakdown.

As my health continually deteriorated, X-rays and tests started showing problems. For instance, one X-ray showed that my colon had completely prolapsed and was laying in my left abdominal area. Another X-ray showed my entire spinal column to have stress points that were critical. X-rays of my lungs showed spots on both lungs. More X-rays showed tuberculosis in both lungs. Don't you think those X-rays showed quite a bit considering it was all "stress?"

If there had been a doctor in the area wishing to practice on something unusual to improve his surgical techniques, believe me I would have been the perfect specimen. I'm sure I could have kept some surgeons busy for quite a while (unless, of course, I expired).

So, naturally, surgery was recommended; in fact, it was called "exploratory surgery." Since I had never had any type of surgery before, I asked the doctors to explain "exploratory surgery." I sat there in awe!

In my case, it meant opening the abdominal wall from just above the pubic area up to the rib cage. That way they could "take everything out that we need to see, cut out what needs removing, and then re-build from there."

I said, "Doctor, you are looking at a girl who is too chicken to have her ears pierced, do you really think I would let you do that to me?" He explained it was my "only chance."

Some of you know the emotional feelings I went through.

First of all, I pleaded with God not to let this be happening to me. When it became clear it *was* happening to me, I blamed God for not giving me the same chance He gave others of my age. After all, I had lived a decent, clean life.

As afraid as I was of the surgery, the desire to live is created so deep in us that I decided if other people had survived that surgery, so could I. My next few conversations with the doctors revealed that even if I had the surgery, there was no guarantee that I would live.

That settled it in my mind: there wasn't any way that I would go through all that and probably still be dead in a few months. My best bet was to forget the surgery, stop going to doctors, and just live on pain pills when things became really bad.

I never told anyone close to me how ill I was. I was embarrassed. Think back 30 years; if one had cancer, gangrene, T.B., etc., you were considered to be dangerous to be around. I can remember my mother would not eat in a particular restaurant because a cook who had worked there five years previously had T.B. As a matter of fact, did you know that 30 years ago, insurance policies didn't even cover cancer treatment unless you bought a separate policy?

I went on with a "no goal" life; doing all the daily things, and just hoping I would not be in pain when I woke up the next day.

It was interesting as I look back. I smoked three packs of cigarettes a day during most of those years I was hunting for my "cure," and never once did a doctor recommend that I stop. In fact, I did stop once, and a doctor recommended that I start again to help me with my "stress."

Like most people today, those people around me then had zillions of bits of advice as to what I needed. If it involved drugs or surgery, it was automatically eliminated from my discussion.

I did start trying to change my diet, though. I substituted honey for white sugar. I used wheat flour instead of white. I slept more and did constructive reading about health and happiness. I tried to do mild exercise.

As time passed, I was introduced to a wonderful doctor who seemed to know from the minute I walked into his office just exactly what I needed. The first thing he told me was that I did not need to die, not right at that time anyway. At first I was mad about that. I had spent months preparing mentally to die, and then this man tells me I can get well.

This doctor was so educated about the body. One thing I liked that he said was, "Drugs and surgery steal from the body, but what we want to do is give to the body."

He explained that if you put in more than you take out, or use, you build a reserve so that in emergencies the body can draw on that reserve for protection. It was so reasonable!

After all the tests, X-rays and examinations he did, it was finally determined that I had quite a few "symptoms" of a total body breakdown. The spots were still visible in my lungs, and even though previous doctors had considered them to be at least pre-cancerous, I decided against a biopsy. My reasoning was that even if they were malignant, I wouldn't have the surgery. So it didn't matter to me enough to have a biopsy.

Both lungs showed tuberculosis. My blood pressure was extremely low, So low, in fact, that during my first two office calls they could not detect blood pressure. I had cysts on my right ovary. I had bleeding stomach ulcers. Previous doctors had me take antacids for the ulcers, which in turn made me constipated, which in turn caused me to have hemorrhoids. My new doctor prescribed cayenne pepper in water with every meal, and between meals whenever I felt indigestion. After one week, I never had trouble with ulcers again. I also have never had hemorrhoids again.

My colon was completely prolapsed and had settled in the area of the left hip. When the doctor pressed around on it, he said the texture indicated complete putrefaction, and he was surprised it was not already in a gangrenous condition. During this time I had a bowel movement every day, and not only is that amazing, but *many* people who have rotten, infested colons also have a bowel movement every day. Therefore they do not realize anything is wrong until it is "too late." The only time I suffered from constipation (as I understood it at that time) was when I was taking antacids.

I have since learned that if you do not have a bowel movement per day for every meal that you eat, you are constipated. If you eat three meals per day, you should have three bowel movements per day. It's that simple. If not, you are constipated. It is interesting how Proper Food Combining corrects that for so many people by automatic regulation of the *whole* digestive tract, not just the colon.

I also had pre-menstrual syndrome (PMS) so badly that I was suidical for at least 10 years. Since it was not until 1980 that the medical society finally recognized PMS as an illness (many individual doctors still don't recognize it), as you can imagine, I was considered to be more than *physically* sick, when one of those tantrums came on.

I had terrible backaches, and my right side hurt if I tried to walk very far or at a fast pace. The doctor also discovered that my liver, pituitary, pancreas, spleen and adrenal glands barely functioned.

My kidneys functioned so poorly that I was slowly developing Bright's disease, and always had swollen, puffy, water-logged legs. Years of deterioration make it necessary for me to take extra precautions with my kidneys to this day.

My gall bladder did not function up to par, and digestion of fats was very uncomfortable, involving hours of bloat, etc. after eating anything with fat in it. When I finally tried a natural means of cleansing the gallbladder, I passed a total of over 500 gallstones on three occasions of cleansing. I had very poor circulation, and of course my hands and feet were always cold. I could not tolerate temperature changes very well. If it was very warm, it was too hot for me. If it was cool, it was cold for me.

My whole lymph system was *very* sluggish. I constantly had a sore throat and swollen glands in my throat. Because of PMS, I had terrible mood swings. *"Nothing"* would happen, and I could go from happy-go-lucky to nearly suicidal in minutes. It was very frustrating and discouraging. I would try to force myself out of a bad mood, but it seemed the more I tried, the deeper down I went.

I suffered from gout for years. If my feet became the least bit cold (even in the summer time), they would swell until it felt like the skin would burst. I could not stand for anything to touch my feet. I would soak them in warm water, increasing the heat in the water until I could feel my feet become warm and they showed some signs of circulation. As the swelling went down my feet would turn black and chunks of flesh would fall off. They would weep fluid for a few days, and then start to heal. In a few weeks the whole process would start over. I have never felt anything so painful.

In those days, I could never remember eating a meal that did not cause me to bloat, burp, experience burning acid, and feel miserable. I would be so hungry, but would try not to eat because I knew of the discomfort it would cause. Along with indigestion, foul gas developed in the whole digestive tract. My problem was that the gas never caused my abdomen to protrude. My abdomen was always flat, but that meant the gas pressed against all internal organs, interfering with their function.

When I finally started eating in such a way that I did not get indigestion I started passing all those gas pockets. I lived on raw juices for almost a year once, and the whole time I passed foul, rotten-smelling gas. (I always said, "If I could bottle this gas, I have enough to put Exxon in second place.")

Of course, my spinal column was not only twisted, making one hip higher than the other, but one hip also was twisted further forward. Years of mis-alignment caused stress on, and interference with, every major organ in my body. I was 50 pounds overweight at my heaviest, and I had cellulite from head to toe.

Now you may wonder why I have filled so many pages explaining my past health problems. In many years of meeting people, and speaking all over the United States, I have discovered that most people have been led to believe that they have two choices to overcome health problems: drugs, and/or surgery!

With all the health problems I had, and there were 28 major ones diagnosed eventually, I *never* used drugs or surgery to overcome them.

First of all, I was too poor to afford the drugs. I was too afraid to have surgery, and my insurance wouldn't have covered the most expensive one, anyway.

In addition, many people have been led to believe that they are not smart enough to figure out what they need to do to maintain their own health. Since I am 'just' an ordinary citizen, and I learned how to care for my body, it gives many people the hope that they can too. We must remember that doctors were not born doctors. They became educated in order to be doctors.

I believe that most of us can become our own 'doctor' in nine out of ten cases. When we can't solve our health problems, *then* we can go to scholars who have learned something we haven't. When you learn about the inside of the body, and how it was designed to function normally, you learn which things to watch. When you feel symptoms, you can 'catch' them much sooner than anyone else, because in effect, you are the only one who senses the problem.

With some of my health problems, I could not correct them myself, so I sought help from persons who did not use drugs or surgery. I have been treated by chiropractors, naturopaths, homeopaths, reflexologists, massage therapists, herbalists, and myself. I have not been treated by a medical doctor for at least 25 years. I am in the best health I have ever experienced, and really feel younger than I did 25 years ago.

Many people who meet me say, "I'll bet you never had a weight problem, did you?" Or, "Have you ever been sick, Lee?"

I give you this background because I want you to see how important Proper Food Combining is. With all these wonderful, educated persons to help me become well, there was still something lacking. I changed from white sugar to honey, I ground my own wheat for flour just before using it. I changed from chocolate to carob. I eliminated coffee and cigarettes.

I still had indigestion and obesity in my life.

One time, at my favorite 'fancy' restaurant, I noticed that if I ate salad, potato, meat and chocolate cake for dessert, I *always* had indigestion. If I left out the cake, I still had indigestion. If I left out either the meat, or the potato, and the cake, I did not have indigestion. At that time I did not know a starch from a fruit, or from a protein. I only knew meat, potatoes and fruit. So, my entry into food combining was very humble and crude. But I *loved* being able to go out for dinner and not get sick. Since I've always preferred starch to meat, I would have a nice green salad and baked potato. One time I had sour cream on my baked potato and developed indigestion. I didn't know then that sour cream was a *protein* fat, rather than just *fat* (as butter is) and that the combination had soured in my stomach.

As my list of facts about foods and how they affected me grew, my health improved. Not only that, I started losing weight, and in the *right places*! I had always had huge thighs, and I started losing them *without exercise*, and without dieting. I didn't know what I had discovered, but I just kept doing the same things day after day out of fear of breaking the spell. Everyone around me started to notice my weight loss and increased stamina.

Since it was common knowledge about my indigestion and health problems, whenever people in my circle of friends developed a 'minor' health problem they would call me. If I had that problem before, I merely told them what I did to get over it. I never realized how many people had indigestion. I always thought I was the only one, but they began to come out of the woodwork once I started eliminating mine.

It is so wonderful to eat and not get sick or fat. I know most people reading this know what I'm talking about. More people are hospitalized in the United States for digestive disorders than for any other illness. It is estimated that 20 million Americans are chronically ill with constant digestive disease. The number one selling drug at the present time is a digestive aid. 200,000 people per year die from digestive disease. $350,000,000 per year is lost in wages due to digestive disease. $90,000,000,000 a year is spent on medical care in this country.

The Number One illness caused by indigestion in this country has to be obesity.

Do you know anyone who is not overweight? It's true that occasionally a person comes forward who wants to *gain* weight, but that is so rare that most people don't take them seriously. And it is not just adults who are overweight: watch people at an amusement park. Even children are suffering from obesity. Their colons are so full of accumulated garbage that their abdomens protrude like their adult friends and relatives.

As my list of foods and the effects kept growing, and I saw more and more health benefits from people changing their diets, I kept hoping someone would come forward with information, like a diet that poor people could afford, so that it would cut down on their medical expenses and lost wages.

In 1977, the U.S. Government released a report that said six out of 10 of the leading causes of death in the United States were due to diet. Do you realize what that means? Not a new miracle surgery, not a new miracle drug, but *diet* could eliminate the majority of the leading causes of death in this country. In fact, at least 100 illnesses have been accredited to digestive disorders and diet.

After reading that government report, I began thinking of ways I could test my food combining on large numbers of other people. I was so thrilled with the changes it made in my life, and I thought it would be so wonderful to help, maybe, 100 other people. (That 100 has since grown to thousands.) Since I've always loved to cook and bake, I opened a restaurant. It was the perfect 'testing ground' because I didn't have money to spend on research nor would I have known really what I was searching for at that time. I was still using food names for my categories rather than food types. I was serving hundreds of people per day and my research and knowledge grew by leaps and bounds.

In the dictionary, *restaurant owners* is listed just a little in front of the word *resurrection*. There's reason for that. I have never worked so hard in all my life. I was the cook (so I'd know for sure what was in the food). I was the waitress (so I'd know who got what). And I was the cleaning lady (so I could afford to pay my bills).

I kept my own records of different "cases." I had people on all kinds of diets. Proper Food Combining always brought better results for health and weight loss. In fact, it was so superior, I never considered any other program to be as effective.

Along with all the hard work, there were wonderful experiences and a depth of knowledge about people and their eating habits gained. For instance, Friday is sort of "eat anything I want day." Almost everyone feels they have worked hard all week, and they are much more apt to have a pastry on Friday than on any other week day. Mondays are strictly a 'diet' day.

Another reason I want you to know my background . . . I've been through it. And even today if I did not use Proper Food Combining and do occasional body cleanses, I would be right back to those old health problems.

You see, that's what I want you to understand. Health is a way of life. The only reason people get sick is from abuse of the body in one way or another. If you want to stay healthy and thin, you must change your way of living. You can't have the garbage and be healthy.

If you were a millionaire and you spent all of your money, would you still be a millionaire? Maybe in your mind, but you'd be out of money. If you "spend" your health, you won't have any left. You may think you're enjoying "spending," but sooner or later you're faced with the truth. The only way a broken millionaire gets his money back is to start all over and work day and night. And that's what you will have to do to regain health.

The difference is . . . it's easier to become a millionaire again because money doesn't age, and we all do. The older your body gets, the harder it is to regain health. *Now* is the time to do something about it.

Please don't be discouraged. Please don't worry because you don't have money. Food Combining costs you less than either the junk food diet or the basic everyday American diet. Many books at the library will teach you about your body and they can be read for free. If you want your own library, purchase my cassette tapes so that you can listen over and over. Remember, I have 25 years of study, research and application as the basis of knowledge. You are not going to learn it and remember it by listening to my tapes one time, or by reading just this book. I have included some actual experiences from some of my clients who are people just like you. They are not miracle cases, they are just average cases. They have put their letters here to help you.

You will notice that many of them mention their weight loss. That is because they are just like you and me. They care about their health, but their main goal through all of this is to be thin and look attractive. Yes, society has brainwashed us to believe that being thin is the most important thing in our lives.

Well, it *is* important. It's just that most people don't know that the way to get thin is to be healthy. At my peak weight I wore size 14 slacks. Today, I wear size 3. That makes me very happy!

I read a nice thought one day that said . . .

If you are planning for a year, sow rice.
If you are planning for a decade, plant trees.
If you are planning for a lifetime, educate a person.

Health is your wealth. Whether it be mentally, physically, or spiritually. You must be balanced in all three to *have it all.*

You *can* have it all, and Proper Food Combining is the start. Feed your body according to its design. It will respond with miracles you never dreamed could happen. Your body is designed to live forever; only imperfection keeps it from doing so. It knows by instinct what to do, if it is given proper care.

Don't just *exist. Live!*

C·H·A·P·T·E·R·2

FOOD AND HOW IT AFFECTS OUR HEALTH

Proper Food Combining

That doesn't sound like a subject we should have to learn. Doesn't it seem like that knowledge should come by instinct, or that we should have learned it somewhere between infancy and adulthood?

We did learn a form of food combining. Remember when you were taught the basic four food groups, and that you should have something from all four groups in your meals? The basic four food group theory is *still* taught in most dietician's schools. Has that type of eating helped or hindered our health? When we see people eating protein, fruit, vegetables and milk products all in the same meal, what outcome can we see with their health and vitality?

In 1980, the United Nations released a report that said that out of 100 nations, the United States was 95th in health. That means that only five nations had worse health *as a nation*. The U.S. is the wealthiest, most prosperous nation in the world, and yet its citizens are sick. That reminds me of the millionaire dying of cancer who was worried about losing his fortune.

At one time, those statistics were not true. Reportedly, in 1900 the United States was the healthiest of all 93 civilized countries. Twenty years later, the United States was *second* out of 93 nations. The U.S. Public Health Service reported in 1978 that the United States had dropped to 79th place.

If we were to check the evolution of processed fast food and drug use, we find a correlation between "modern day miracle discoveries" and the deterioration of the average person's health. As a matter of fact, many "modern day miracles" have contributed toward this deterioration, but most of them cannot be changed in our daily lives by our individual actions.

One thing we *can* change though, is what goes into our own mouths. Now that may sound easy enough, but in reality you must remember we are facing a highly-commercialized opponent when we want to simplify our eating.

Companies which sell processed foods spend millions of dollars to make your taste buds prefer processed foods over "natural" foods. So, if you are to choose between a fresh apple and a piece of pastry, don't be surprised if you prefer the pastry. The "system" is against you! However, you *can* change your eating desires to the extent that you never even try the pastry, and that will eliminate making a choice.

Fats

Every day, millions of people eat foods that have been specifically proven to be harmful. Much exposure relating to the fast food industry is showing that people are eating things they would not have believed. For instance, in eating french fries, many people who are vegetarians were surprised to learn that those french fries were cooked in *beef fat*. As a matter of fact, the Center for Science in the Public Interest says, "Most people don't realize that the biggest fast-food restaurants fry foods — even apple pies — in 90 percent or more beef fat."

Beef fat is loaded with saturated fatty acids. It increases blood cholesterol levels and the risk of heart attacks. Since statistics show that one out of five men, and one out of 17 women, will have a heart attack before they are 60 years old, it would be worth a person's time to analyze the eating habits perpetuated by fast food restaurants.

USA Today recently contained a report on the fat contents of foods produced by 8 different restaurant chains:

- Roy Rogers breakfast crescent sandwich with ham — 42 grams of fat
- Burger King Whopper with cheese, french fries and vanilla shake — 67 grams
- Hardee's large french fries — 17 grams
- Hardee's cheeseburger, french fries, soda — 30 grams
- McDonald's Big Mac, french fries, milk shake — 53 grams
- Arby's deluxe potato — 38 grams
- Wendy's Double Cheeseburger, french fries, Frosty Dairy Dessert — 75 grams
- McDonald's Chicken McNuggets — 19 grams

Since 50 grams of fat (or less) per day is recommended to protect against heart disease, one can see that one stop at a fast food restaurant could easily exceed this recommended maximum daily fat intake. A U.S. Government survey of the average American diet (the Second National Health and Nutrition Examination) shows that hamburger is the largest single source of fat in the American diet.

Because so many people suffer from heart disease, one form of which is arteriosclerosis, or "clogged arteries," caused by the *wrong form of fat*, many people think they should eliminate all fat from their diet. It is interesting to note that in Alaska, the "traditional" Eskimo (that means one who has not learned modern ways), eats three to five pounds of tallow a day in the *raw form*. In fact, 90 percent of the traditional Eskimo's diet is *fat* and yet they *do not* have a cholesterol problem in their arteries. It is not just the amount of fat eaten that is important. It is what the body does with it.

Free Radicals

Items known as *"free radicals,"* attack the healthy cells in the artery lining, puncture into the muscle sheath, and cause swelling. Free radicals interfere with normal cell reproduction and cause build-up on the artery wall, which creates bumps, and that is the beginning of the hardening of the arteries.

Some causes of free radicals are chlorine, refined sugar, soft water, smoking, poor food combining which putrifies and sends carbon monoxide into the blood stream, industrial chemicals, oral contraceptives, hydrogenated oil, rancid unhydrogenated oil, caffein, food additives, radiation, gas, constipation, fumes from autos, and many other detrimental modern day wonders.

Cholesterol, or fat, is not the root cause of America's health problems. Only fats that the body cannot *use* build on an existing problem. Eliminate *all* of the causes of free radicals, and you will see a significant difference in cholesterol buildup.

Diet Can and Does Cause Illness

Have you ever suffered from indigestion, heartburn, tiredness, aching joints, headaches, mood swings, sugar cravings, constipation, allergies, asthma, hayfever, obesity, hypoglycemia, diabetes, salt cravings, high blood pressure, yeast infection, continual colds or sinus infections?

A Mayo Clinic newsletter made the following observations:

"Yes, it is estimated that approximately one-half of the United States population is medically classified as chronically ill. $400 billion is spent annually on health research, and approximately $8 billion is spent on prevention. It is also estimated that 90 percent of all illness that people bring to doctors is either self-limiting or beyond the medical profession's capabilities for cure."

In the 30 years that I have been interested in health, it has been easy to see that a major health problem in this country is *obesity*. I have sat at amusement parks for hours, counting and watching people. I have analyzed their body shapes to estimate the location of their weight problems. I have watched them eat: the ingredients they eat, the speed with which they eat, how long they chew their food. It has been an interesting study in that people vary only slightly. I have followed them around after watching them eat to see if they have any energy, if they burp, if they eventually take an antacid, if they go to the bathroom.

It is a vicious circle: a sick body craves sick food. Sick food creates sick bodies. A healthy body does not crave junk. It loves fresh fruits and vegetables, and whole grains.

You will notice if you sit and watch, how many people, even if they are thin, will have a bloated-looking stomach.

Health is basically the responsibility of each of us. There are some communicable diseases which may affect us, but if we learn all that we can about how the body functions, "according to its design," we can even protect ourselves against them to a degree.

It is very interesting to me that most people will eat what they want, (even when they know from many sources that the item is bad for them). They will practice many things that are known to be harmful to the body (such as smoking, overeating, overdrinking, and so on). Yet when they hear of a health program which is made up of all the good things that they know to be beneficial to the body, they say they are afraid to try it "because they don't know if it is dangerous or not."

One time I had a customer who was on 19 drugs. Even her doctor had told her to find some way to get off the drugs because of the side effects. When I discussed with her the issue of changing her eating habits (her present ones were atrocious), she became very skeptical and thought it might prove harmful in her case.

This woman had been constipated her whole life. When she went on some herbs and increased the raw foods in her diet, she started having approximately three bowel movements per day. She lost 19 pounds in a month due to all the old rotten feces that was coming out every day.

Her blood pressure normalized (something the drugs never helped accomplish in 17 years). Her doctor took her off three drugs: a digestive aid, a diuretic, and blood pressure medicine because she no longer needed them. However, she found eating properly "just too hard." She went back to her old ways, regained the weight, and went back on the drugs.

Sugars and Sweeteners

The American public eats 291,000 *tons* of sugar per year. They also consume 32,249 *tons* of artificial sweeteners.

First of all, the name 'artificial' sweetener should give us a clue. Second, what results do we find from these artificial sweeteners?

Why do you think most people use them? To lose weight. (There are some diabetics and others with health problems, but the majority use them to cut down calories). However, do you know anyone who uses artificial sweeteners, who no longer has a weight problem? Have they really eliminated their weight problem? No.

When I had my restaurant I noticed that most people who used them had just as many pounds to lose as those who didn't. Therefore, why the substitute?

There has always been a controversy surrounding artificial sweeteners. Years ago my doctor told me not to use them because of the constant stress they put on the kidneys. He also mentioned that people who drink a lot of sodas with artificial sweeteners had a 60 percent higher chance of developing cancer of the bladder. In addition, he said that, in effect, they are a drug. Saccaharin was proven to have carcinogenic effects. Now the latest is aspartame. This sweetener is used in most of the 20 billion cans of diet and soft drinks consumed each year by Americans.

Tests on humans fed 1.8 grams of aspartame per day (there are 1.9 grams in a six-pack of diet soda), showed that participants reported headaches, cramps, weight gain (isn't that ironic?) and menstrual irregularities. It is thought that these symptoms are from the methyl alcohol contained in the aspartame.

Some countries have warning labels that children, pregnant women and those with neurological disorders should not take products which contain aspartame. The rest of us get no warning against neurological disfunctions, infertility, impairment of vision, seizures, headaches and poor circulation, all symptoms of aspartame's side effects. By the way, symptoms of methyl alcohol poisoning are identical to the symptoms of multiple sclerosis. And while the scientists, businessmen, and others are bickering back and forth over this, let me tell you a couple of experiences my students have related to me.

A woman, 60 years old, started drinking sodas sweetened with aspartame (NutraSweet). She developed headaches, which she never had before. She had dizzy spells. Then she had little "blackout" spells when she would lose her thoughts for a few seconds. Then she started having longer "blackout" spells. Medical examinations showed nothing, but when she casually mentioned she had started drinking diet sodas for the first time in her life, because it "had the new safe sweetener," her doctor asked her to quit drinking them for a while. Within two weeks, all "blackouts" stopped, and within three weeks the headaches were gone. It was the *only* change she made.

A woman, 35 years old, started using all NutraSweet products for herself and her children. They all suffered headaches and general "run down" feeling. By chance, she saw an article about NutraSweet and the suspicious methyl alcohol poisoning potential. Elimination of NutraSweet products also eliminated all headaches and 'run down' feelings. Is sugar better?

How to Make Refined Sugar

From Beets: The total process is long and complicated. First, slice the beets. Then treat with very hot water until all sugar is extracted in the form of diffusion juice. Next, add *lime* and *carbonic acid* (derived from a lime kiln) through the juice until the lime is converted into a *carbonate* (a salt of carbonic acid), which carries away the mineral elements and other "impurities" of the juice. Pass the juice through a filter press which will evaporate the juice. *Bleach* it to give it a pure white color.

From Cane: After the juice is extracted, mix with *lime* and pass through the fumes of *burning sulphur*, wash with salts of *tin* or *bluing* in a centrifuge. Use various forms of *acid phosphates* (salts of phosphoric acid) to clarify the cane juices. *Bleach* again and follow with evaporation. The remaining solids are sugar. Granulate these solids. The finished product is more than 99 percent sugar. It is an unbalanced food, completely devoid of all minerals, vitamins and other accessory nutrients normally found in natural foods.

After you have gained this product, it will help you achieve the following health results: Tooth Decay — Malnutrition — Obesity — Diabetes — Low Blood Sugar — Hardening of the Arteries — Heart Disease — Emotional Imbalance — and many, many other things.

Milk

As soon as some mothers see on the food combining chart that milk should be consumed "alone," they panic. They can't imagine their children eliminating milk from their diet, or drinking it alone.

I heard an herbalist speaking on one occasion, who had been a doctor when he served in the military during World War II. He was on a ship, and he said the men on the ship were suffering from allergies, colds, and other respiratory ailments, all the time. He finally had milk eliminated from the crew's diet for 2 weeks. He estimated that 90 percent of the allergies and colds were eliminated as a result. It was so successful that milk became "taboo" for the rest of his tour of duty.

Even though it is true that antibiotics are used on cows and then the potential of passing that to humans is there. Even though it is proven that *all* animal life on earth is usually weaned by the age of three. Even though cow's milk has a completely different makeup because it is designed to help cows grow, not humans, and thus humans do not have a digestive tract designed to digest cow's milk. Even with all this, the real issue is what health benefits do you get from milk. *Or* what damage?

Humans are the only animals that drink milk after they are weaned. In addition, humans go to a different animal type for their milk. They cross over from human to cow or goat. No other animal would do that.

If your children repeatedly become ill with 'flu, colds or similar ailments and you take them to the doctor, did you ever notice what he does? The doctor removes milk from their diet becuase it causes mucous.

Consider that. If your child is drinking milk and that milk is causing mucous and colds and infections, what is your child's problem? The milk. That's why doctors often take the child off milk. But then, as soon as your child starts improving, the doctor has you resume the milk — the very thing that caused the problem.

When we remove milk from a person's diet, we *always* see some type of health improvement. Most people can tolerate a fermented milk product, but not regular milk.

Since we have been brainwashed to believe that we must have the calcium from milk (and that is the "only" reason to drink it), it seems logical to find a food to use instead of milk that would provide that calcium without the side effects of milk. That food is Lee Tee — a combination of dried barley leaf and wheat grass juice.

On a weight comparison basis, (milligrams per 100 grams) milk has 100 milligrams of calcium, while Lee Tee has 1,108 milligrams. Now there's a source of calcium. Not only does Lee Tee *not* contain indigestible fat like cow's milk, but it does contain many other vitamins and minerals which the body can assimilate very easily.

As a matter of fact, there are various foods such as sesame seeds which provide an excellent source of calcium (and other necessary minerals), but that don't have a lobbyist as strong as the milk industry. When milk is fermented, approximately 50 percent of the milk sugar is broken down, and a supply of lactase (the enzyme required to digest lactose) is generated in the fermenting process. Fermented milk generates a "friendly bacteria" which is known to help the whole digestive tract eliminate "unfriendly bacteria."

In my whole life, I have probably consumed 10 glasses of milk. I do not suffer from osteoporosis, or bone degeneration, or rotten teeth. Along with eliminating milk from my diet, I also eliminated refined sugar, red meat, pork, preservatives, food coloring, coffee and refined salt.

You see, health is not just one food item. It is a total program of eating properly.

Additives and Preservatives

Food and Drug Administration regulations only require the actual listing by name of one food color. This is the food color known as yellow dye No. 5. That one is listed because so many people are allergic to it. All other food colors are listed simply as "artificial color."

There are over 2,000 artificial flavors on the market. Only a few are required by law to be noted on the label. The average American eats one teaspoon of artificial colors, flavors, and preservatives per day. That's almost four *pounds* per year.

Many foods do not require labeling, so we don't actually know what is in them. Actually, manufacturers have the option of noting all ingredients or just some of them. Some foods which do not require complete labeling are chocolate, mayonnaise, ice cream, milk, salad dressing, preserves, processed cheese, cheese, canned fruits and tomato products.

Also, some ploys used by the food industry may fool us only for a while. Have you ever seen the words, "No preservatives added"? Didn't you think to yourself, "Oh, good. There are no preservatives in this!" Wrong.

You see, a product can come to a manufacturer with a preservative *already* in it. But if the manufacturer doesn't *add* a preservative, the package may still legally read "no preservatives added." The preservative in this case is called a "secondary additive" and doesn't count. It counts to me, how about you?

All of this doesn't even include sulfites. We will discuss them later.

Meat

The use of lots of flesh protein has been stressed so much, that most people feel they must have meat in their diet every day or they will become deficient in protein. There are many sources of protein besides meat, and because of the use of antibiotics in the raising of animals, it makes one wonder, "How safe is meat?'

Studies estimate that about half of the 35 million pounds of U.S. produced antibiotics are given to animals and the other half prescribed for humans.

According to research from the Federal Center for Disease Control, evidence has demonstrated conclusively, for the first time, that feeding antibiotics to beef and dairy cattle, hogs, and poultry, breeds a novel form of microbe that can later infect humans.

An excellent source of protein is raw nuts. Many people do not realize that vegetables, fruits, and grains contain considerable protein, but must be eaten properly, at the right time, in order to be digested and be of benefit.

Raw Foods — Cooked Food — Canned Foods — Frozen Foods

Sometimes it amazes me that the human race is even still existing when I see how processing has taken over our fruits, vegetables and grains.

After being grown on soil that is nearly depleted of natural ingredients, our food is loaded into bushel baskets and taken to a plant to be frozen or canned at once. Publicity tries to convince us that everything is "quick frozen" so that all of the minerals and vitamins are maintained. The average frozen food changes temperature 17 times. With each change nutrients are lost. The truth is that when foods are canned and heated, most nutrients are destroyed.

For instance, consider the case of green beans. Three-fourths of the vitamin C is lost during canning. In fact over 50 percent of all the major nutrients in green beans are lost during canning. In addition we have to allow for some loss during storage before the canning took place. Many think that freezing is much better than canning because there is no use of heat and it is done faster. However, more and more information is becoming available to indicate that freezing changes the molecular structure of food.

In raw foods, minerals are bonded in an organic form to enzymes, proteins, amino acids and sugars inside each cell. You have heard them by the name *chelated minerals*. However, when foods are heated or frozen, different acids and compounds intrude upon these mineral bonds, converting the minerals into an inorganic state which is not easily absorbed by humans. So, even though the minerals are not "lost" in the heating or cooking, they are "lost" as far as the body's ability to assimilate them. With freezing, the practice of "blanching" is usually necessary in order to inactivate enzyme systems and "enzymes" are necessary for digestion.

Changing the form of a food makes a large difference too. For instance, apple juice (fresh) is ingested more than 10 times faster than the juice still in the whole apple.

Applesauce is ingested more than three times faster than the whole apple. What is interesting here is that the body handles the apple differently, not only when the fiber is completely removed, but also when it is merely physically disrupted, as in cooking it into applesauce. Also, when ingesting applesauce and apple juice, there is a rebound fall in blood sugar levels that does not occur when eating the whole apple. This is due to the fact that there is a higher level of insulin in the blood after consuming juice and sauce, than after consuming whole apples.

Remember the saying, "An apple a day keeps the doctor away?" Besides containing that valuable "pectin," apples are a delicious fruit that can usually be purchased all year round.

Almost all foods in their natural raw state are low in sodium and high in potassium. The introduction of agriculture and food processing increased the amount of sodium and decreased the amount of potassium. Persons with cancer, AIDS, yeast infection, high blood pressure, strokes, heart attacks and kidney diseases are almost always deficient in potassium.

The main thing I want you to consider is this: try to eat all your food raw, or as plain as possible. The more seasoning, sauces, etc. that you use, the more food categories you use. This is very hard on our bodies, because we are designed to eat fruits, vegetables and herbs. In the beginning food never came "sweet and sour", or "Benedict," or any other way, except raw.

Microwave

I don't think all of the information is in about microwave cooking, but there's enough in for me to never use one.

It's convenient, it's fast, and sometimes the food looks better than other forms of cooking. However, research over the years continues to show conflicting results, and, as they say, "Where there's smoke, there's fire." Continued use may not cause cancer in humans as it does in rats in experiments, but it does change the molecular structure of food, and that affects assimilation of the nutrients, and that's wrong enough.

I have met numerous people who became ill after eating food cooked in microwave ovens, and they told me they "always" get ill if they eat something cooked in a microwave.

Instant Foods

Researchers in Australia have found that instant foods produce a higher rise in blood sugar than unprocessed foods. When they examined digestibility and blood sugar response, the starches provided oral bacteria with fermentable sugar, which turned into acid that eats tooth enamel and causes tooth decay.

The foods tested were instant rice, cornflakes, rice cereal, corn chips and instant potatoes. All of them produced a rise in blood sugar, except the potato chips, and it was felt that their high fat content caused slower absorption (*American Journal of Clinical Nutrition,* December 1985).

Drinks

Simple sugar consumption provides over 53 percent of Americans' carbohydrate intake, and soft drinks make up a large percentage of that. Because these drinks also contain other addictive ingredients, it is hard for some people to break the soft drink habit. Once the habit is broken, however, the soft drinks taste "too sweet."

There is nothing that can replace freshly squeezed juices. They contain natural distilled water along with nutrients. Raw fresh juices contain enzymes, which enable the body to digest food, and to absorb it into the blood.

You cannot have life without enzymes. Once you kill the enzymes in food, you kill the life in it. That is why raw juices are better than canned or bottled juices. Enzymes can be frozen, but they cannot be heated above 130 degrees without being killed.

A juicer machine, which does or does not extract the juice, is quite a financial investment. (They cost between $100 and $500.) The money you save on health costs, plus the nourishment your family receives from it, will more than pay for the initial cost. I strongly recommend *everyone* own a juicer, and *use it.*

Salt

The national research council says the daily intake of sodium, safe for adults, is 1100 to 3300 milligrams a day. Most Americans consume 5000 to 7000 milligrams a day. One teaspoon of salt contains 2400 milligrams of sodium.

When buying products low in salt there are various terms used which can be confusing, such as:

Sodium Free: Less than 5 milligrams per serving.

Very Low Sodium: 35 milligrams or less per serving.

Low Sodium: 140 milligrams or less per serving.

Reduced sodium: Processed to reduce the usual level of sodium by 75 percent.

Unsalted: Processed without salt, whereas the food normally is processed with salt.

Salt can be hidden. Check every label if you are still using foods in cans, boxes, or jars. Over 55 percent of the American diet is processed foods, and they have the highest salt content of all. They also do the most damage to the body.

Vegetables in their natural form, contain sodium in its balanced form, and the body uses it well. Need I say it again: "Nothing beats the foods that were designed to match the body."

Herb Teas

This is a subject after my own heart. Herbs are a natural health aid and a natural food. All nations on earth have used herbs as food, poultices, inhalants, and salves. Did you know that recent findings indicate herbs have electrical energy flows, which match different organs of the human body, and can therefore help to charge those organs to regain health?

Herb teas are an excellent drink, but many people get discouraged changing over to them because the tastes seem "blah." I have found that by making the tea stronger than the recipe suggests, plus drinking one that has my favorite flavor, I enjoy them more than when worrying about the health problem they are going to solve. Most people like mint, and it really is an enjoyable tea, especially on a cold night. Also, licorice is enjoyable hot or cold. Check your local health food store for different flavors.

If you add honey, molasses, malt, sugar, lemon or anything else to tea when you drink it, it becomes a meal. You must allow at least one hour for digestion time. Plus, you do not drink with meals, if you want to digest them, so either drink the tea plain, between meals, or have it as a meal, with sweetening.

"Uppers"

Many people literally cannot "think straight" until they have had their morning dose of caffein.

Did you know some people actually suffer withdrawals similar to drug withdrawals, when they try to abstain from coffee? They suffer "the shakes," terrible headaches and nausea. They become irritable, have no energy, and cannot control their emotions.

Some people do not use coffee for their "upper;" some people use soda drinks or chocolate. Both contain caffein. Some people use other "mild" drugs.

Why not test yourself? Try to go without whatever your "upper" happens to be. Now, don't start on a workday, start on a day you are going to be home alone. Pick a day when things are not too demanding and you can afford to feel lousy. If you last for two weeks and then go back to your source, you will experience a surprise at the first "fix."

If you were an addict, you will be grateful to have broken the habit, because many underlying health problems stem from "uppers."

One more point: If your children needed an "upper" to make it through the day, every day, what would you think?

Water

Most people realize that if they did not bathe every day they would soon have body odor and cleanliness problems. However, many people do not realize that the inside of their body needs the same kind of bath in order to help rid the body of odor and uncleanliness internally.

All fruits and vegetables contain distilled water. All of the vitamins and minerals in the water in the fruits and vegetables are absorbed and can be used by the body. Either within the fiber of the fruit and vegetables, or in the juice, we have an excellent source of food and distilled water.

However, since juice is a food and has to be eaten as a meal, it is important that we have an additional source of water, as well. The controversy over water continues. Some say if you don't drink water with minerals in it, you will develop a mineral deficiency. Other persons say if you do drink water with minerals, you will develop mineral deposits. Therefore, after years of following persons living both ways. I can see without a doubt that:

1. tap water can kill you;
2. water with lots of minerals *does* leave deposits in the joints and other parts of the body;
3. in order for your body to get a bath internally, the water should be either steam distilled, reverse osmosis, or some method used to remove hard minerals, additives and impurities. And
4. never drink tap water where a water softener is used. It is high in salt.

When people start my program, they usually increase their water intake. If you increase drinking water that is high in salt, you may gain weight, instead of losing it, because the salt will cause water retention.

Getting Started

I have found that if people are really serious about losing weight and gaining health, they pay attention to a total program. For instance, haven't you heard how bad refined sugar is for you? Haven't you heard it rots your teeth, shatters your nerves, upsets digestion, makes you fat, destroys vitamins and is just plain bad for you? We have all been hearing that for years. And yet what do we find? Most people still eat refined sugar.

I'm sure most of you have tried many kinds of diets — especially if you have a weight problem (and who doesn't?) Proper Food Combining is not a "diet" diet.

Proper Food Combining is the proper way of eating for which the body was originally designed. It is like "returning to nature." I hope you will try combining for at least one week. When you return to non-food combining, you will understand what I mean.

We each have an important decision to make: for some of us it is simply "too hard" to be healthy. For others, the effort becomes a joy as we begin to see the signs of radical improvement in our health, attitudes and enjoyment of life.

The choice is yours. The next chapter offers a complete introduction to the lessons I have gained over 30 years of experimenting and research. They can change your life and your outlook, and all you have to do is learn to control what goes into your mouth.

C·H·A·P·T·E·R·3

PROPER FOOD COMBINING

In 1977, the U.S. Government released a report which showed that six out of 10 causes of death in the United States are due to diet. (Select Committee on Nutrition and Human Needs — Press Conference, held Friday, January 14, 1977, in Room 457, Dirksen Senate Office Building.) That means that not drugs, not surgery and not the latest miracle treatment, but diet, would eliminate all those illnesses.

It is also known that 90 percent of all adults in the United States are overweight, caused in part by indigestion. Five billion dollars are spent each year on antacids. The number one selling drug in the United States is Tagamet, a digestive drug. The number one reason people are hospitalized in the United States is the treatment of digestive disorders. The facts pretty well show that we must do something to change our diet in order to aid our digestion.

Have you ever been on a diet? Are you overweight?

Many people today are eating very good diets and yet they'll say to me, "Honestly, all I live on is fresh fruits and vegetables, and I'm still overweight and feel terrible." The problem is, they're mixing the wrong foods together, which, as a result, is causing a toxic condition inside the body in the digestive areas. Or they are preparing their food in such a way as to cause a toxic reaction. And that toxic reaction is causing the same damage as though they had eaten junk food.

Our Original Diet

Let's look back, for a moment, at the human body's origin and our original diet.

In the Bible, in Genesis 1:29, it says, "and God went on to say, here I have given to you all vegetation-bearing food which is on the surface of the whole earth, and every tree on which there is the fruit of a tree bearing seed. To you, let it serve as food."

You notice it's saying that at the very beginning of Man, his food was supposed to be vegetation and fruit. These foods were specifically designed for the human body. These fruits and vegetables, and herbs would have, no doubt, been ripe when picked, and they would have been eaten right away. When foods are eaten right after harvest, they contain all the nutrients that are lost by freezing, cooking, canning, etc. That alone makes them easier to digest because of the live enzymes in them.

If we look back into the history of the ancient Israelite nation, we find that they were considered to be an outstandingly healthy nation among all the other nations. They practiced many things which made them more healthy than the other people around them. And if we take a close look at their history, and their commands regarding food, we find they were actually instructed by the Creator of the body regarding not only *what* foods to eat, but how to prepare them using sanitary precautions.

One instance of the latter is that instructions are given in the Bible that no blood was to be eaten of any animal or fowl. Flesh was allowable, but no blood. Today, we understand this as a health law because the blood is where sickness is (Leviticus 17:14).

There's something else I'd like to mention here. The prophets, the ancient men of old, practiced something called fasting (Joel 1:14). In fact, before they were to make some important decision, or have a big, drastic thing happen in their lives, they would fast.

You might stop and think, "Why in the world would people go without food when they're going to make a big decision?" Today, most people would think that's the very time you should eat. They think they should try to build their energy, increase their strength. But you see, believing that way shows that we do not understand the value of letting the body cleanse and take care of itself.

Eating food only gives energy and strength when it is fully digested *and* assimilated. Have you ever been so distraught or upset over something that you could not eat? Or maybe you did eat, and then developed indigestion because you ate too fast, or were just under too much stress.

Overeating and improper food combining is practiced so strongly today that people really don't understand the value of going without eating. There really *is* a value in going without eating.

The ancients knew that it gave the body a time to relax and rest. The mind and the body both would not be worrying about having indigestion, having to go to the bathroom, having to burp or having a sour stomach. They wouldn't be tired after having a heavy meal. I'm sure we've all experienced that. It takes a lot of energy to digest a meal, especially a large meal.

After fasting, they would be in a restful state in which the mind and body would be able to cleanse and relax and have a rest before making a heavy decision.

Food and Energy

The way to get energy and health from food is not only by putting it in our mouths, but by chewing it; digesting it at each stage of our digestive channel; assimilating the nutrients, and then expelling the waste on time. From the mouth to the rectum, timing is of the utmost importance. When the food is in the mouth, it's important how long it's there. When the food is in the stomach, it's important how long it's there. When it's in the small intestine and in the colon, the timing's important there, too.

There are various reasons why, but firstly, it's because in the mouth, stomach, small intestine and colon, digestion of various food types takes place.

One thing we must establish: the human digestive tract was never designed to digest complex meals. Thanksgiving and Christmas Day dinners are perfect examples of the worst meals you will ever eat. With those meals, you have everything wrong. You have protein, starch, fat, sugar, refined sugar, fruit and for many people, alcohol, all thrown in together.

Have you ever seen so many people take an afternoon walk as they do in the afternoon on Christmas and Thanksgiving Day? And we all know why, don't we? They all have indigestion from overeating. They are taking the walk to "wear off that big dinner." Are you kidding? It would take a month of walks to wear off those meals!

Besides the body not being able to digest complex meals, the body also cannot digest non-food items. It is designed to digest food only.

So first of all, let's understand what real food is and then we can proceed. Food is any material which can be incorporated into and become a part of the cells and fluids of the body. Now keep that in mind. It must be able to be incorporated into and become part of the cells and fluids of the body.

To be true food, it must not contain useless or harmful ingredients. As an example, non-useful drug materials are all poisonous. For instance, consider aspirin. These materials leave a residue in the body because the body cannot digest it and incorporate it into the cells. Food can have *undigestible* parts for roughage, but not un-usable ingredients.

Not all raw plants are foods because some contain poisons. Therefore, we can't even say that all raw plants are foods. In the raw state, foods contain proteins, carbohydrates, fats, mineral salts and vitamins and enzymes. Some have more and some have less indigestible roughage.

If food reaches a certain digestive organ and the food is not in the digestive condition it should be, it will not continue to digest. Let me repeat that. If food reaches a digestive organ, and the food is not in the right condition of digestion, it will not finish digesting. In fact, it will putrefy.

On page 44, you will find a diagram of the digestive tract. Please refer to that now. For food to be usable, it has to be broken down into minute particles. At each section of the digestive tract, there are glands producing enzymes which set up the action for this to occur. The mouth, the stomach, the small and large intestines and pancreas are all capable of producing enzymes of the most minute combinations if (1) they are not overtaxed; (2) we do not make wrong combinations; (3) we do not overeat, and (4) if we do not eat non-food items. If we do any of these things, whatever food we eat (and I include true food) turns to poisons. The poisons sit in the digestive tract long enough to not only damage it, but also the toxic residue can seep into the bloodstream and body tissue. Eventually we can see it on our bodies as cellulite, obesity, and disease.

Diets

Most of us don't really worry about indigestion and that backup of putrifying food until we see it as cellulite or obesity. And then we really begin to worry.

For years we eat, we get indigestion, acid stomach, have allergies, asthma, bronchitis, headaches, backaches, swollen limbs, varicose veins, gallbladder trouble, liver trouble, eczema and all kinds of other problems to which we adapt. Then, all of a sudden one day, we notice we have ugly fat. *Then* all of a sudden we become very worried about our health, how we should eat, and how we can get rid of it.

Imagine, most of us cope with illness, but let us start to look bad and we really panic. Vanity steps in and says to us, "You'd better do something about helping that poor old body." In fact, if most of us are honest about it, we are much more concerned about our appearance physically than our health. We'll go to great length, even beyond reason sometimes, to lose weight and get rid of cellulite on a much more serious basis than we would to eat the right foods to make sure our health is good.

I've talked to many women whose calorie intake is less than 1000 calories a day. They're starving themselves trying to maintain their weight. And for some of them, it's not just for beauty. They also want to be healthy. But with poor food combining, even a small amount of food will not digest. It putrefies at different sections of the digestive tract, and the end result is that these people gain weight. These same wome have doubled their calorie intake with Proper Food Combining, then lost weight.

Allergies

For many years we've heard the word allergy. Everyone seems to have allergies. Allergy is basically non-digested protein poisoning. Normal digestion delivers nutrients; indigestion delivers toxic poisons.

For years, medical science refused to believe that allergies were connected to diet. Now more and more research is headed in that direction. Allergies are a sign of toxic poisoning in the body and *can* be from many sources, but I have found when people eliminate indigestion, their allergies usually leave also.

Indigestion

So here's another basic. When you've eaten a meal and you get indigestion, know for sure that that meal is turning to poison and the rest of your body is going to have to pay for it. I constantly have people tell me, "Every time I eat I get indigestion." Well, shouldn't that tell you that something is wrong? To detect indigestion, the food need only go from the mouth to the stomach. So it has to be somewhere between those two parts of the body. Most people think it is the body that's malfunctioning. They never imagine it's what they're putting into it.

Three Step Check

In order to identify your wrong combinations, you have three clues which speak the truth to you. They cannot be wrong, so listen to them.

(1) You will eat the combination and get indigestion.

(2) You will step on the scale the next morning, and it will have gone up.
(The proper way to weigh is . . . get up every morning, urinate, and weigh naked. Weigh the same way, at the same time of the day, every day).

(3) You will have bad breath, from both ends, the morning after. When the food ferments, it ferments the whole digestive tract. It is possible to not know you had indigestion the day before, because it may have been mild enough you didn't notice it. It is impossible to not know when you have the bad breath . . . you will taste it, smell it, or someone will tell you about it.

Pay attention to these clues. If you have one, two, or all three clues, you have combined wrong, or you ate a meal too soon after your previous meal.

Learn to read the inside of your body. These clues can be more important than an X-ray, and considerably safer..

Amounts and Combinations

Many times strict dieters don't think it's the food they're eating because they're eating such a small amount. They'll say, "All I had was an apple and a piece of chicken." But an apple and a piece of chicken eaten together interfere with each other's digestion.

As a matter of fact, I had a young woman come to me who was approximately 20 pounds overweight, and as usual, her extra weight was from the waist down. She had been living on 600 calories per day for over six months, and she had *gained* seven pounds. Her pancreas was painful and she had indigestion and hunger pains all the time. She had very bad pre-menstrual syndrome, and absolutely no energy.

Her doctor had put her on a high protein, low carbohydrate, no fat diet, with a 600 calories per day limit. When she didn't lose, but actually gained weight, he presumed she was "cheating."

She ate a small piece of baked chicken, or fish, three times a day. She ate one piece of fruit per day. She gradually changed her diet to the exact opposite of her high protein diet over the next three months, and she eventually was able to eat all she wanted of any one food type at a time. Her pancreas improved, her extra pounds came off, in the right places, and her mental attitude improved phenomenally. She no longer counted calories or carbohydrates. Instead she counted raw enzymes from fresh fruits and vegetables, and eliminated dead protein from her diet completely.

Four years later, she continues to do well as long as she practices Proper Food Combining. Several months off the combining due to a family crisis sent her back to her old eating habits *and* old health problems and weight gain. Once she returned to food combining, and regained her health, she said, "The world can have its diet, I'll never leave food combining again."

If you watch animals in their natural habitat, they eat one kind of food at a time. They don't order a seven-course meal. They don't have a brain like man to reason and to build a Thanksgiving Day dinner, which will kill them. Rather, they eat by instinct, and instinct tells them that for the best, easiest digestion, they should eat one food at a time.

Here's a little side point that's interesting, too. I have found that when humans are very healthy, they desire to eat in a healthy way because their instincts work better and so does their brain. They automatically want the foods that are right for the body. I used to crave white sugar and chocolate, and would have bet my life I'd never see the day I could refuse either one. They have both been replaced with honey and carob long ago.

Enzymes

If we combine foods right, don't overeat and eat real food, our bodies have something called enzymes which act as catalysts to break the food down to nourish us. Substances which do not ordinarily combine with each other when brought together may be made to do so by a third substance when it is brought in contact with them. This third substance does not in any way enter into the combination, but its mere presence brings about a combination and reaction. This third substance is called a catalyst. And the catalyst in the case of food digestion is enzymes. Much research is in progress regarding enzymes. So far, it is estimated there are over 3,000 known enzymes which feed our bodies.

Enzymes cause the food to become something that it wouldn't ordinarily be. Enzymes cause a reaction in the food but the enzyme itself does not actually become part of the food. The action of enzymes closely resembles fermentation, but it's not the same as fermentation, because fermentation is caused by bacteria and causes toxin and putrefaction, whereas enzymes cause food to produce nutrients.

Each enzyme has its own action and it only acts on one specific food type. Enzymes that act on protein do not act on fats or sugars or starches. To be even more specific, within each food type, there are specialized enzymes to act upon their variety. For instance, among the sugars there are maltose, sucrose, lactose, etc. Each one of these sugars has its own enzyme and one enzyme cannot digest the other's type.

It is important to note that *heat* destroys enzymes. Since enzymes are the ingredients in food that help the food digest itself, you can see the reason 70 to 80 percent of your food should be eaten raw. Once the enzymes are destroyed, you have ruined a good part of the food's nutrients.

Digesting Starch

Enzymes also must act in order. They can perform their work *only* if the preceding work has been done properly by the enzyme that precedes it. For instance, starch. We should eat any starch dry, as dry as possible, because digestion of starch starts in the mouth. The saliva in the mouth secretes an enzyme called ptyalin, and this enzyme converts starches into the sugar form called maltose. A major part of starch digestion must start in the mouth. If you eat your starches dry and

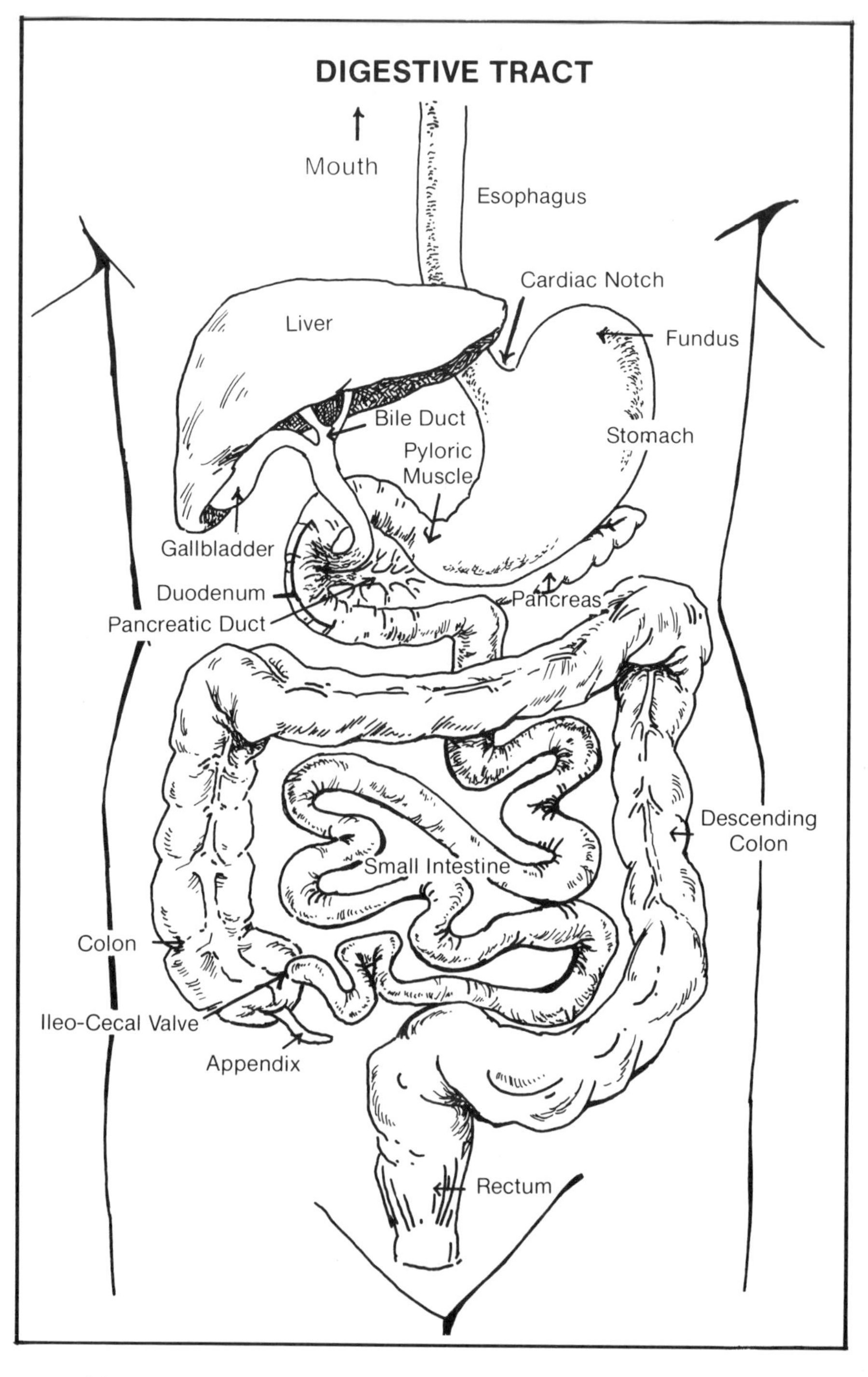
DIGESTIVE TRACT
Mouth
Esophagus
Cardiac Notch
Liver
Fundus
Bile Duct
Stomach
Pyloric
Muscle
Gallbladder
Duodenum
Pancreas
Pancreatic Duct
Descending
Colon
Small Intestine
Colon
Ileo-Cecal Valve
Appendix
Rectum

chew them until they are in a liquid state when you swallow them, you will be sure you have enough ptyalin in your saliva to start the conversion of starch into maltose.

In the stomach, starch will continue to digest if it's in the proper condition, and if it has enough ptyalin. It may continue to digest for two to three hours. It then leaves the stomach for the first part of the small intestine called the duodenum. As it leaves the stomach, it passes the pancreas.

The pancreas is very valuable for its digestive enzymes, too, and it secretes amylase, which is similar to ptyalin. It helps the next step of digestion of starch. When the starch reaches the small intestine, an enzyme called maltase will break the maltose, formerly starch, into the sugar glucose. However, the maltase in the small intestine cannot convert the maltose into glucose *unless the starch has already been converted to maltose* in the mouth. Starch must be in the form of maltose before it reaches the small intestine in order for the small intestine enzyme maltase to break the starch down further into the sugar glucose.

You must understand how important it is that each section of the digestive tract carries its own load and performs its own job. You cannot change that; it is in the design of the body. Please refer to the diagram on the opposite page to review that part.

The mouth converts starch to maltose. The stomach churns and acts sort of like a treatment plant. The small intestine converts maltose into glucose. It must be in that order. Ptyalin in the mouth cannot convert maltose into glucose. Neither can maltase in the small intestine convert starch into maltose.

Refer to your digestive tract picture again. Notice that from the mouth to the rectum, each section has its duties. If that digestive system were not abused by over eating, combining poorly, and eating 'junk' food, it would last "forever." Each section would carry a little of the load.

You might ask, "What happens to it if the food's not in the right condition at the right time and place?"

If it is not in the right condition, it cannot be turned into glucose, which is where you get your energy. In addition, it will become toxic because it cannot be used, and it putrefies.

So now we have the process for starch. You put starch in your mouth, the enzyme ptyalin breaks it down into maltose. When the maltose goes into the stomach, it continues digesting. As the food goes into the small intestine, it will be converted into glucose by the small intestine enzyme maltase. Finally, in the form of glucose, it will be used by the body as energy. Foods must be converted to the sugar glucose in order for the body to use them as energy.

Starch Blockers

To show how important it is not to interfere with the digestive process, let's use the example of starch blockers.

Starch blockers are made from a specific substance from grains and beans which will prevent the amylase from the pancreas from working. Therefore, some modern-day scientists decided that if you couldn't digest the starch, or finish the digestion, your body couldn't gain the calories from it. Also, if it doesn't digest the starch, you cannot then produce the sugars maltose or glucose. They thought this would keep your blood sugar from increasing, and it would reduce the body's tendency to manufacture fats in the blood from glucose. Therefore, they decided that this would help diabetics, plus keep you from gaining weight.

They presumed diabetics would be helped because the end result would be there would be less sugar in the blood from the digestion of starches.

Some interesting research was conducted by Dr. Jeffrey Bland at the Bellevue-Redmond Medical Laboratory, in Washington. The following are some of the results of that research.

First of all, it was proven it would take 100 to 200 mg. of starch blockers to reduce amylase by 50 to 100 percent. Lower doses didn't do anything, which means that almost all starch blockers on the market wouldn't work sufficiently because they contain less than 200 mg.

But let's suppose you took the blocker that was strong enough. Can you decipher what would happen? What do you think happens to undigested starch when it reaches the colon? When bacteria come in contact with something like that, they are capable of fermenting the undigested starch and producing gas and bowel toxins.

Tests have shown that there was considerable increased breath hydrogen gas produced after administering high-potency starch blockers. This indicated increased intestinal fermentation of the undigested carbohydrates.

Remember, it was hoped that this would help diabetics. When this was tested on animals, it led to enlargement of the pancreas and potential increased risk of pancreatitis because when the starch left the stomach, the pancreas sent out amylase, but the blocker kept blocking the amylase from digesting the starch. So the pancreas sent out more amylase, and more amylase, and the pancreas was overworking itself trying to compensate for the amylase inhibition by secreting more amylase.

Damage of the pancreas, is already the problem of diabetics. So when the pancreas exhausted itself by trying to overcome the starch blockers, the damage to the pancreas was even worse.

Now how do you help a diabetic by damaging and overworking the pancreas? Many health food stores used to carry starch blockers. Many of them are still fighting to sell them because now blockers have been banned by the government. So, let this be a little warning — not everything sold in health food stores is good for you. You need to obtain the knowledge of the inside of your body so that you yourself can discern what you should or should not consume.

You see, the world knows that people want to lose weight and they'll try anything at any cost. If people had understood the digestive process and had understood what the blockers do, do you think people would have bought them? It's true that since the ban on starch blockers went into effect, they are no longer on the market. But what if that ban were lifted? Would you take them, just because they are legal?

Digesting Protein

When protein is fully digested, it becomes known as amino acids.

Protein is not digested in the mouth like starch. Protein digestion starts in the stomach. In the stomach, the gastric juice produced by the stomach contains an enzyme called pepsin. And there's a strong acid that it produces called hydrochloric acid. Hydrochloric acid and pepsin instigate protein digestion.

It is interesting that even though protein digestion initiates in the stomach with enzymes, the initiation of enzymes actually starts in the brain when you think about eating protein.

Then at different stages of digestion, different protein-splitting enzymes take over. For instance, the stomach enzyme pepsin converts protein into what are called peptids. That happens in the stomach. When it leaves the stomach, it passes the pancreas and receives a shot of the enzyme trypsin. It then goes to the small intestine, and in the small intestinal fluids, there is an enzyme called erepsin. Erepsin finishes the conversion of peptids into amino acids.

However, erepsin cannot work directly upon complex protein. So that means if the protein went directly from the mouth into the small intestine, the erepsin could not do anything to that protein because it has not yet been converted to peptids, the step that would have been done back in the stomach. Without prior action of pepsin, erepsin would not act upon the protein food.

I want to add something very important at this point. Have you ever burped and felt it burn? You know how you feel it come back up into the esophagus?

Between the esophagus and the stomach, there is a muscle called the cardiac notch that closes once the food has gone from the esophagus into the stomach. Now that's a safety feature for you. That muscle is there so that once food is in the stomach and it starts digesting, it cannot come back up into the esophagus.

This is most important when protein is digesting. Remember, I mentioned that hydrochloric acid and pepsin mix together to digest proteins. This mixture is very, very strong. In fact, it's like corrosives; it can burn flesh. Remember, *its* job is to break flesh down so it's in a condition in which it can be assimilated by the body.

You might be wondering. "If it's that strong, why doesn't it burn holes in my stomach?"

The stomach is the only part of the digestive tract

that produces a mucous lining to protect itself from that strong acid. Therefore, as long as the food is mixed with that acid, it must remain within the stomach or it will burn the digestive tract.

Indigestion causes such combustion that it forces the cardiac notch to open, and the digestive foods mixed with acid spit up into the esophagus. It burns the esophagus because there is no protective lining there. I have heard a discussion among certain medical doctors as to the cause of cancer of the esophagus, and it's from that very thing: that constant burping, that constant burning; bringing that food back into the esophagus where it should not return.

Let's review that. First you chew the protein real well in the mouth. Then it goes to the stomach. The protein starts digesting with pepsin and hydrochloric acid in the stomach. The pepsin and hydrochloric acid break the protein into peptids. The peptids leave the stomach and go to the small intestine, where an intestinal juice and pancreatic juice reduces peptid into amino acids. If pepsin has not converted protein into peptids before leaving the stomach, the erepsin in the small intestine cannot convert it into amino acids.

Some of these words are so similar. As we're going along, you may want to turn to the illustration which shows the digestive tract on page 44 and write the names of the enzymes next to the proper organs as well as what food the organs digest. Then you have a ready reference showing you exactly what each section of the digestive tract does.

Previously we learned that protein must have acid to be broken down. An alkaline medium, ice-cold drinks or desserts, and wrong foods stop the action of hydrochloric acid and pepsin. So if any one or all of these three things is mixed with protein when you eat it, you stop the digestive action. If you don't have pepsin and hydrochloric acid, you don't digest the protein. If the protein leaves the stomach undigested, it cannot be digested any place else in the digestive tract and it turns to toxins. One result is allergies due to undigested protein.

Ice-cold "anything" gives the body a shock. It especially affects protein digestion because protein needs strong enzymes. If the stomach is shocked by something cold, it can take hours for the stomach to return to normal in order to continue producing these enzymes.

I used to think if I ate something cold and held it in my mouth for a few minutes, it would be warm by the time it reached my stomach. We must remember that the whole digestive tract is connected together, and that it sends messages from one part to the other to protect itself.

So when the cold is in your mouth, it affects the stomach and the esophagus. And when it reaches the stomach, it shocks it even more. Since protein digestive aids are manufactured in the stomach, it stops production. And it stops digestion until the body can return to normal. This could be hours. Thus, the food in the stomach lies there much longer than it should and, of course, starts to putrefy.

I'd like to add an important point here. If you drink liquids with your meals, no matter what you're eating, you dilute the stomach enzymes and acids. When these enzymes are diluted, they are not strong enough to complete the digestion in the stomach.

At the other end of your stomach, there is a muscle, the pyloric muscle, which controls the amount of food leaving the stomach. It does not all leave at the same time.

Because the food should be broken down to liquid form when leaving the stomach, the muscle will start allowing liquids to pass — only liquids. The liquids you drank with your meal leave first and since the liquid you drank has a percentage of the diluted digestive aids, it takes those along when leaving the stomach. This causes a load on your digestion. The stomach would either have to re-produce enzymes for that meal, or the food would leave as soon as it became liquid and it simply would not digest because of the missing, already expelled, enzymes.

Digesting Fats

The gastric juices of the stomach lining produce an enzyme called lipase, and that enzyme splits fat. Hydrochloric acid also helps digest fats in the stomach.

When the fats leave the stomach, bile from the liver and pancreatic juice from the pancreas produce enzymes which help to digest fats. If you'll notice in the diagram on page 44, it is convenient that they're all located there so that as the fats leave the stomach, the bile from the liver is handy, as is the gallbladder. The pancreas is also right at that location. So all these different types of enzymes can be added effectively to the food to do whatever else needs to be done.

Eventually, fat is converted into what is called fatty acids, which then are broken down to glucose. It is then ready for assimilation into the body for use as energy. Fats take the longest to digest of any of the foods. Remember, any time you eat fat with any food, you use the longest digestive time to figure your time. For example, if you eat fat, a 12-hour digesting food, and you eat it with starch, a five-hour digesting food, your total digestion time is 12 hours (for the fat, not five hours for the starch).

Digesting Non-Starch Vegetables

Non-starch vegetable digestion is different from starch vegetable digestion because they don't need large amounts of ptyalin or a strong alkaline base in order to be digested.

Even though I refer to them as *non-starch*, they do contain some starch, but in a much smaller quantity than a starch vegetable. They also contain protein (a proportion similar to the amount of starch) and fat (a very small amount). Since the starch, protein and fat are all from the vegetable kingdom, and because they are in smaller, more balanced amounts, they are compatible with either starch or protein.

Scientists are still amazed at the variance in strength and timing in the production of digestive enzymes, and they still are not exactly sure when or how it happens. It is known that for every food kingdom, and at every stage of digestion, the body automatically knows which enzyme to produce and how strong to make it.

The body only makes enzymes strong if *needed*. The body does not waste energy.

Non-starch vegetables (either raw or cooked, but raw is preferable) should be chewed very well in the mouth and should be in a liquid condition, from saliva, before being swallowed. You will not need ptyalin because the stomach produces the enzymes for non-starch vegetables.

When the non-starch vegetable reaches the stomach, enzymes are produced to digest the starch, protein and fat. Since this is not an animal protein or fat, the enzymes required will not be of the strong acid potency.

Non-starch vegetables will spend approximately five hours digesting in the stomach and will be in a liquid condition when leaving the stomach. As they pass the pancreas, they will receive any necessary enzymes for the next step of digestion. In the small intestine they will be converted into glucose sugar.

Let's clarify the reasons non-starch vegetables can be eaten with starch *or* protein.

Remember — acid and alkaline conditions *do not mix*. The body will not produce digestive aids for them at the same time. Therefore, if you were to eat broccoli (a non-starch), with baked potato (a starch) that is okay. Even though the potato needs a strong *alkaline* base, that does not interfere with the broccoli because the broccoli does not need a strong *acid* base.

If you were to eat broccoli with baked chicken (a protein flesh) the same principle applies. Even though the chicken needs a strong *acid* base, it does not interfere with the broccoli because the broccoli does not need a strong *alkaline* base.

Note: This also explains why even *mild*-starch vegetables will not digest with protein. Mild-starch, like starch, needs the stronger *alkaline* base.

Digesting Fruit

Since fruit is one of the foods designed especially for the body, it has to be good for you, right? Well, it is, and yet many people eliminate it from their diets because they think it is too high in sugar, or find they develop a bloated gas condition when they eat it, or develop bacteria growth as with candida.

The digestive compartment designed to initiate digestion of fruits is the small intestine. If you will refer to the diagram on page 44, you will see that the small intestine is beyond the mouth and stomach area. This, of course, means that the fruit must pass through the mouth and stomach in order to reach the small intestine.

Fruits have a digestive time requirement of two to three hours, depending on the category. If you eat fruit on an empty digestive tract, it will be sent to the small intestine very fast. When it reaches the small intestine, different enzymes digest it, gradually converting it to glucose sugar.

Problems develop when fruits are eaten with other foods, or at a time when the digestive tract is *not* empty.

If you will refer to the digestive time requirements on your food combining chart (pages 68 to 71) you will see that proteins and fats require 12 hours digestion time, starches five hours, and fruit two to three hours.

Let's assume that you decide to go on a very *good* diet. You eat broiled fish, green salad, and an apple for dessert. That meal will ferment just as if you had eaten a pastry for dessert. This is the reason. The broiled fish and green salad are going to take 12 hours to digest. That means the apple is also going to take 12 hours because it will be in the stomach along with the fish and green salad.

Do you know what happens if you grate an apple and put it in a warm, humid environment for 12 hours? It starts to ferment. And that is exactly what happens when you eat an apple, or any other fruit, at the end of a meal. Fruits must be eaten alone. And each category of fruit must be eaten alone.

If you change your eating habits with fruit, you will find that it is very good for your bowel movements. It will give you energy instead of fermentation, and "according to its design" it will help to cleanse your body.

It is the fermenting of fruit which gives you the bloated gas and abnormal bacteria growth in the digestive tract, *not* the fruit!

Absorption of Nutrients

Years ago I thought everything that had to do with digestion happened in the stomach. I was not aware of the different digestive compartments. What a relief and thrill it was to find out about the marvelous organization and design of the digestive tract, from the mouth to the rectum.

Once food is properly digested, it is important that the nutrients from it be absorbed, because after the conversion of food to glucose sugar (our only source of energy) or to amino acids (our cell builders) our lives depend on it.

This is a very deep study in itself and is covered in my cassette album, *Gaining Health/Losing Weight*, but I want to mention that nutrients are absorbed by the body through the wall of the small intestine. The intestinal wall contains minute blood and lymph capillaries which lead to larger lymph and blood capillaries via the Mesentery Wall. These capillaries eventually carry nutrients to the liver, blood stream, and cells. You must realize that if those capillaries can carry nutrients, they can also carry toxins and putrefaction. And that is exactly what happens.

Last Meal of the Day

Many families do not eat their last daily meal until six o'clock in the evening, or later. Then, even after that, they will snack while watching television. This is a very dangerous habit.

A large meal reaches peak digestion approximately seven hours after consumption. Try to make adjustments in your eating time, retiring time, and size of evening meal. Large dinners are not necessary and proper adjustments will aid your sleep, your next day's mood, and your next day's vitality.

Sometimes even when combining properly, you may still get indigestion. That could be because your previous meal did not have enough time to finish digesting. It could be you ate so late before retiring, that your meal did not digest at all. Once you lie down, your whole body rests and digestion almost stops. This total body rest is necessary for health.

The danger in this is . . . When a meal does not digest properly, it causes fermentation in the stomach. That fermentation expands, causing pressure against the heart. People have died in their sleep from this very thing.

To digest your evening meal, eat as early as possible.

Food Combining: What NOT to Do

Using your chart of the different food combining groups, let's examine the different food groups, and why we would not eat them together.

We have one food type called sweet fruit; another called syrups and sugars; another, acid fruit; another, melon; and another sub-acid fruits. Then there are starches and mild starches, which are grouped together the same for food combining. There is a non-starch/green vegetable category. Notice that protein is divided into three groups: protein flesh, protein fat and protein starch.

Digestion takes place basically in the mouth, stomach and small intestine. At the different locations, enzymes are secreted for different food types. In the mouth, enzymes start starch digestion. In the stomach, enzymes start protein, non-starch vegetables, and fat digestion. And in the small intestine, enzymes start fruit and sugar digestion.

The body knows its limits, and it will not produce

enzymes when certain other enzymes are being produced. Therefore, I have divided all foods into groups which show you which ones will or will not hinder each other's digestion.

Here are some examples of how to use the chart on pages 68 to 71 for combining. Each food category is separated from other categories. Notice the pages that list fruits. All of the categories on those pages must be eaten alone. Knowing that saves you having to turn the pages back and forth.

The other pages list categories that *may* be combined with other categories. Notice the number within each square. The number represents the number of hours needed to digest that food. That means, when you eat that food, you should wait that many hours before you eat again.

At the bottom of each square you will see the words, "eat with." That will explain which categories you may combine with the one you're reading. You cannot combine acid and starch together. Acid, like tomato, inhibits the production of ptyalin, which digests starch. So, an acid food and a starch food cannot be digested together. Refer to your chart and find the starch category. Notice that it does not list acid fruit as a combiner. Also find acid fruit on your chart and notice that it does not include starch as a combiner.

Protein and starch. Hydrochloric acid inhibits ptyalin production, the enzyme for digesting starch. In fact, large amounts of hydrochloric acid destroy the ptyalin enzyme completely. Therefore, if you ate protein, which is digested by hydrochloric acid, with starch, which is digested by ptyalin, which the hydrochloric acid destroys, how are you going to digest the starch? It's impossible. Starch requires an alkaline condition for digestion and protein must have an acid condition. The two are incompatible.

Then you may think, "Well, protein is protein. I can probably eat all the proteins together."

But your chart shows that even proteins of different categories cannot be mixed together. Different enzymes are required for the digestion of different proteins, and they each mix differently with other food types. It is possible that you could eat two types of flesh, but you would not mix the different kinds of food types in the protein and eat them together.

For instance, you might eat steak and hamburger together. But you would not eat hamburger and cheese (protein fat) together. They're two different kinds of protein. They require different kinds of enzymes, and they mix with different kinds of food. Protein starch would not be eaten with either protein flesh or protein fat. Remember, too, nuts, (protein fat) should always be eaten raw.

How about acid and protein? Gastric juice from the stomach lining will not pour out for protein digestion if there is acid food in the mouth or in the stomach. By inhibiting the flow of gastric juices for that digestion, you hamper the protein digestion and that results in putrefaction.

There does seem to be an exception to the rule for some people. Refer to your food chart and find acid fruit. Notice where it says, "Eat with — maybe protein fat"? (A full explanation of this can be found on page 44 of the *Proper Food Combining Cookbook*). Some people can digest protein fat with certain acid fruits. Remember, I said protein *fat* not all proteins. Just protein fat with certain acid fruits.

You must check each one for yourself. Some people cannot eat any acid fruit with protein fat combinations. But fruit acid and protein fat, such as cheese, nuts, avocados, etc. may be eaten together by some people.

I mentioned the fact that fats always delay gastric juices and digestion. Acid food also delays gastric juices and digestion. So since they both do it, some people can eat these together and all it merely means is that it takes two, three, four or five hours longer for their meal to digest. But if they wait long enough, their systems, for some reason, will digest it.

Now, consider fat and protein combinations. The category fat, for instance, oil or butter, and the category protein would not normally be eaten together. This is because oil delays the secretion of digestive juices, making the protein lie there longer. So it would only be eaten when you eat protein fat, that fat would be okay to combine with protein.

It's interesting to note that if you eat an abundance of green vegetables, and especially raw cabbage, it counteracts the inhibiting effect of fat. So if you eat a lot of green vegetables, say a big green salad, along with having cheese (a fat protein) or oil dressing, the delaying of the digestion is not near what it would be otherwise. There's something about those raw, green vegetables, which science does not even understand yet, that keeps the fat from interfering with digestion.

Another poor combination is sugar and protein. Sugar inhibits the secretion of gastric juice in the stomach, which digests protein. Remember, you have to have a strong digestive acid, like hydrochloric acid, to digest protein. But sugar stops the secretion of that acid.

Another thing, sugar digests only in the small intestine. So sugar itself should be sent there very quickly on an empty stomach. If it isn't and is delayed, it lies in the stomach and will ferment. Then when it gets to the small intestine where it should digest, it's already fermented.

I'll give you an example of that. Let's take sweet and sour chicken, sweet and sour spareribs. You chew them in the mouth, but nothing happens to them there except saliva mixes with them. They get to the stomach, the stomach starts trying to shoot out enzymes to digest the protein, but the problem is, there's sugar in the food. The sugar prevents the stomach from producing those enzymes properly. It lies there and it churns and it churns and it churns and it takes about twice as long to break down as it normally would. By the time it leaves the stomach, the sugar has fermented, because sugar lying in a nice, warm place for approximately 10 to 12 hours ferments and becomes alcohol.

So, when all this mixture leaves the stomach to go to the small intestine, the sugar is fermented. The protein is probably only half digested. It all gets down to the small intestine, and the small intestine is where the nutrients are supposed to go back into the system. But what do we have? We have fermented sugar, which will send toxins back into the bloodstream and the tissue.

How about sugar and starch? No. The very same things result. Sugars ferment while they're waiting for the starch to digest. Also, it is believed that sugar interferes with the starch digestion. The mouth will produce saliva but there will be no ptyalin in it. If there's no ptyalin, the starch that you ate with the sugar cannot be digested.

Just a little side thought here. We read so much in the news today about the immune system, the lymph system, being toxic. All major illnesses being discussed now, such as AIDS, yeast infection, leukemia and cancer, are all affected by the immune system. One of the major reasons for toxic lymph is the putrefaction of sugar in the body. Sugar that is not digested and ferments puts a very, very taxing load on the lymph system.

Another category is milk. Milk is a half-protein, half-fat combination food already. I do not recommend using milk, but if you have it, drink it alone as a meal. And if you look on your digestive chart, you wil see how long it takes for milk to digest. After you have consumed milk alone, you should wait 12 hours before you eat again.

We've learned three foods that delay digestive secretion in the stomach: acid, sugar and fats. Here are some other causes. Cold desserts, cold drinks, anything iced or real cold while you're eating a meal, will interfere with the stomach producing digestive juices, and, as we already discussed, liquids with a meal will also interfere.

Let's briefly review the food combining groups that you should *not* use together.

You do not eat acid and starch. You do not eat protein and starch. You do not mix your proteins. The three different protein categories that you see on your chart do not mix together. You do not mix acid and protein flesh. You do not mix acid and protein starch. For some people, acid and protein fat are permissible occasionally, but you must check to make sure you do not get indigestion. You do not eat fat with protein flesh. You do not eat sugar with protein, *period.*

Do you realize what no sugar and protein together means? No cookies, no cake, no pastries. You do not eat sugar and starch and there we go again. No cookies, no cake.

At this point you may be saying. "Wait a minute. Cake is only one food if I eat it alone. Or if I just have cookies, that's only one food."

But let's analyze the ingredients. They have flour, a starch. They have eggs, a protein. They have sugar, a sugar. They usually have nuts, raisins, chocolate chips, or coconut; and there's four more categories. You see, we may call it one food, one piece of cake, but it contains up to 10 different food categories. That's why pastries are always fattening. Within themselves they contain wrong combining.

Don't be misled like I was for years. I had decided. "No more goodies from the grocery store. I'll make all my own from good ingredients."

I ground my own flour fresh just before baking, and it was wheat flour mixed with wheat germ. I used eggs from the farmer next door who had roosters in with the hens. I used honey which had been drained from the comb instead of sugar. I used baking powder from the health food store so that there was no aluminum in it. I used fresh shelled walnuts, sunflower seeds, carob chips, and unsweetened coconut to make them extra healthy.

And you know what? I gained weight and had indigestion from them just as I did from the chocolate pinwheels from the grocery stores. You see it's not just a matter of good ingredients. It is the *combining* that makes the final difference.

When food types interfere with each other's digestion it can cause all the ailments listed at the very beginning of this chapter. The list included indigestion, acid stomach, allergies, asthma, bronchitis, headaches, backaches, swollen limbs, varicose veins, gallbladder trouble, liver touble, and finally, obesity and toxic fat.

Poor food combining meals are setting up a toxic reaction. That toxic reaction is then sent through the whole digestive tract. All the way through it leaves a residue of poisons that back up through the tissue, through the lining wall of the small intestine and the colon. These poisons then enter the bloodstream and the tissue, via the lymph and blood capillaries.

To show you the inter-connection between your whole digestive tract, if food is in the stomach and it putrefies, the stomach sends a message to the colon that says, "Secrete a mucous protective lining because I am sending something to you that you will not believe."

Just think, if the food is toxic in the stomach, what will it be like when it reaches the colon, perhaps as long as 18 hours later. The body is so aware of the damage from putrefied food, the protective mucous lining in the colon remains there approximately 72 hours to allow plenty of time for all that toxin and its residue to be removed from the body.

Proper Food Combining: What to Eat Alone

Now let's use our food combining chart as a worksheet. Go to the category of melons and see the word "alone"?

I want to point out something about melons. Even in its own category, melons should not be mixed together. If you eat cantaloupe, you should eat cantaloupe, not cantaloupe and watermelon; not honeydew and watermelon. You will find that when you mix melons, even though they're in the same food group, they all have an enzyme type of their own. I have many people say to me, "Lee, I can't eat melons. I know watermelon's good for flushing out your kidneys and I know there's a lot of distilled water in melons, but every time I eat them, I get terrible gas."

You know why? Because you're mixing them. I have eaten half of a watermelon before without bloating and it would do a lot of flushing of the kidneys. The water and nutrients in melons help remove salt from the tissue and it and the pulp create a tremendous bowel movement. Melons are an excellent food, but you cannot eat them with other foods or even other melons.

The faster a food digests, the faster it spoils. So when you look at your food chart and you see the digestive time of the different foods, realize which ones are going to spoil the fastest if you combine them with something else.

Melons digest very fast. It takes different enzymes for each one, and if you slow the digestion of one of them, it can spoil very fast. Eat them alone.

In the other groups mentioned, syrups and sugars, sweet fruits, sub-acid fruits and acid fruits, are eaten alone. However, sometimes you can eat more than one particular fruit in the same group. Some people can't eat two together, even from the same category. If you can, it does make a delicious meal.

In the acid fruits category, some people can eat orange and grapefruit together, and some people like pineapple and tomato together. Other people cannot eat combinations even within the same food type. You're going to have to do a little experimenting with that to find out whether you can or not.

As we learned previously, fruit is digested in the small intestine. If anything is in its way between the mouth and the small intestine, the fruit will get stopped along the way, and while it lies there waiting for something else to be digested, it will then ferment and turn into a poison.

The goal in learning Proper Food Combining is to eliminate fermentation of food. It's that fermenting that is setting up the toxic buildup that sends poisons back into the body.

We want to be able to eat food and have it digest as

quickly and easily as possible. Then the nutrients will pass through the lining of the intestine back into the bloodstream, into the liver, into the lymph, wherever it's supposed to go, and we will receive nourishment from the foods, rather than toxins.

Water and Eating

Without water your body cannot survive. It is important that you have water every day, and for most people that should probably be about 64 ounces, or eight glasses.

If you train your body, you will be thirsty at the proper time. For instance you may be drinking water with your meals. That will definitely hinder your digestion. In fact, drinking anything with your meals will hinder your digestion.

But you are probably thirsty during meal time, and you may even think you won't be able to swallow the food unless you drink with your meals. Liquid dilutes your digestive enzymes. Also, since the stomach releases liquid after a meal is eaten, it will release the water, which will contain the diluted enzymes, and the bulk of your meal will be left in the stomach without proper enzymes. The stomach will either have to manufacture more enzymes, or the food will not be digested properly. Either way, the food will be in the stomach a delayed period of time.

The proper time to drink water is first thing in the morning on an empty stomach. This will in effect "bathe" the digestive tract. You should drink water at least 15 minutes before eating a meal. You should wait approximately two hours after you eat before drinking water again.

At first this may seem difficult to you, but it is interesting that once you practice this for a while, you will wonder how you ever drank with your meals.

Once you are not drinking with your meals, you will notice that you will not have that "bloated" feeling after you eat. You will also chew your food better and longer when you do not drink, and that extra chewing will help your food digest better.

If you drink plenty of water between meals, you will not be thirsty when you eat. Once you do not drink with your meals, you *will* be thirsty in between meals.

Never drink ice cold water as that shocks the digestive tract. Never drink very hot water, either; around body temperature is best, no matter what you are drinking.

Proper Food Combining: What to Combine

Now let's discuss the other food groups. Find the column headed "starches."

Starches and mild starch are combined with the same food. You have a category of starches and a category mild starch foods, like beets and carrots. They can be treated in the same way as potatoes, banana squash, pumpkin, etc. Those foods can be eaten with fats and green vegetables. Notice on the chart at the bottom of each category it says, "Eat with . . . fat and/or non-starch/green vegetables."

Next is the non-starch and green vegetables category. Notice at the bottom of the list it says, "You can eat these with fat, starch, and mild starch *or* (notice that's *or*, not *and*) your choice of protein flesh, protein fat or protein starch. Remember that's *"or"* not *"and."*

I mentioned before that some people's digestion is good enough that they can digest acid fruit with the items in the category of protein fat. Some people can eat tomatoes with cheese. Also, some people can tolerate the acid fruits, tomatoes and lemons, with non-starch vegetables. If they can handle that, they can have those two categories with the non-starch and green

PROPER FOOD COMBINING CHART

★DIGESTION TIME IN HOURS

NON-STARCH/GREEN VEGETABLES

Artichoke
Asparagus
Bamboo Shoot
Bell Pepper
Beet Top
Bok Choy
Broccoli
Brussel Sprout
Cabbage
Cauliflower
Celery
Chive
Collard
Chard
Cucumber
Dandelion
Eggplant
Endive
Escarole
Garlic
Green Bean
Kale
Leek
Lettuce
Mushroom
Okra
Onion
Parsley
Peas, fresh
Radish
Spinach
Sprouts
Squash (except Banana, Hubbard)
Swiss Chard
Turnip Top
Watercress
Zucchini

EAT WITH:
Fat, Starch, Mild Starch or (choice of one) Protein Flesh, Protein Fat, Protein Starch or Raw Tomato, Lemon

★5

STARCH

Bread
Cereal
Chestnut
Corn
Cracker
Grains
Jerusalem Artichoke
Lima Bean, fresh
Pasta
Peanut, raw
Popcorn
Potato
Pumpkin
Rice
Squash (Banana, Hubbard)
Yam

EAT WITH:
Fat, Non-Starch/Green Vegetables

★5

PROTEIN STARCH

Beans, dry
Peas, dry
Soy Beans
All Soy Products

EAT WITH:
Green Vegetables

★12

MILD STARCH

Beet
Caladium Root
Carrot
Parsnip
Rutabaga
Salsify
Turnip

EAT WITH:
Non-Starch/Green Vegetables, Fat

★5

PROTEIN FLESH

Beef
Chicken
Duck
Egg
Fish
Goose
Lamb
Pork
Rabbit
Seafood
Turkey
Veal

EAT WITH:
Non-Starch/Green Vegetables

★12

PROTEIN FAT

Avocado
Cheese
Kefir Cheese
Nuts (except Chestnut, Peanut), Raw
Olives
Seeds
Sour Cream
Yogurt (full fat)

EAT WITH:
Fat, Non-Starch/Green Vegetables or Fruit, Acid (maybe)

★12

FAT

Butter
Cream
Margarine
Oil

EAT WITH:
Starch, Non-Starch, Mild Starch or Protein Fat

★12

Chart continues on next page

FRUIT, ACID

Acerola Cherry
Apple, sour
Cranberry
Currant
Gooseberry
Grapefruit
Grape, sour
Kumquat
Lemon
Lime
Loganberry
Orange
Peach, sour
Pineapple
Plum, sour
Pomegranate
Strawberry
Tangelo
Tangerine
Tomato

EAT ALONE or maybe with Protein Fat. Or maybe, Lemon and Tomato with Non-Starch/Green Vegetables

★2

FRUIT, SUB-ACID

Apple, sweet
Apricot
Blackberry
Blueberry
Boysenberry
Cherry, sweet
Elderberry
Fig, fresh
Guava
Huckleberry
Kiwi
Mulberry
Nectarine
Peach, sweet
Pear
Plum
Prickly Pear
Quince
Raspberry

EAT ALONE

★2

FRUIT, SWEET-DRIED

Apricot
Banana
Date
Fig
Peach
Pear
Pineapple
Prune
Raisin

EAT ALONE

★3

FRUIT, SWEET-FRESH

Banana
Black Currant
Mango
Muscat Grape
Papaya
Persimmon
Thompson Grape

EAT ALONE ★3

MILK

EAT ALONE ★12

MELON

Cantaloupe
Casaba
Christmas.
Crenshaw
Honeydew
Muskmelon
Nutmeg
Persian
Pie
Watermelon

EAT EACH MELON ALONE ★2

SYRUP, SUGAR

Brown Sugar
Carob
Honey
Malt
Maple Syrup
Milk Sugar
Molasses
White Sugar

EAT ALONE ★2

vegetables. But they cannot add protein flesh, such as eggs and flesh. So, if a person wants to have a green salad with tomato and cheese on it, that's permissible. But you cannot add eggs, flesh, tuna, or anything like that. If you want to have a salad with tuna, that's permissible, but you have to leave out the tomato and the cheese.

It's important to make sure you understand this. With non-starch/green vegetables, you can have a starch such as potato. You can have a mild starch such as beets, and you can have fats such as oil. OR you can have non-starch/green vegetables with protein fat such as cheese if you can digest it. OR you can have protein flesh or protein starch with just non-starch/green vegetables.

Next, let's discuss the category of fats. Fats such as oil, butter, cream, etc. may be eaten with non-starch/green vegetables, starch and mild starch *or* protein fat and non-starch/green vegetables. Any fat you eat should be unhydrogenated.

Hydrogenation is important to understand. In order to maintain the shelf life of oils, scientists came up with a process known as hydrogenation. They blend hydrogen into the oils and the purpose is to keep the oil from fermenting, or turning toxic or rancid in the bottle.

We hear a lot about fats as they are associated with high blood cholesterol and arteriosclerosis, (or hardening of the arteries). The *saturated* fats are the ones that receive the blame for these two illnesses, so many people will eat only *unsaturated* fats, in an effort to eliminate high blood pressure and hardening of the arteries.

The problem is that unsaturated fats tend to "break down easily," and in order to stabilize them, they are "hydrogenated." During hydrogenation, fatty acids are changed, making the oil undigestible because it is no longer in a natural state. This fatty acid change has been linked to such illnesses as pre-menstrual syndrome, pancreatic disease, breast and pancreatic cancer, heart attack, arteriosclerosis, etc.

Start checking the labels when you shop for baked goods; you will notice they all say "hydrogenated oil". Some companies have tried to outsmart us by using the term "partially hydrogenated." It is the same thing as far as digestion is concerned. Hydrogenation turns liquid oils into solid margarine, and when you think about how much margarine this country's population must consume, you get an idea of the problem.

Oils which are not hydrogenated are sold in health foods stores and called "cold pressed." If you eat fats, it is best to stick with cold processed oils or butter, though it is probably a good idea to reduce your use of them too.

Current information shows that once oils are removed from the seeds or nut, it is in an abnormal condition even if it is unhydrogenated, or cold pressed. Try using the whole raw nut or avocado for your oil source.

Another problem is that when you eat hydrogenated oil, the body cannot break the fat down to assimilate it and use it, so it turns rancid in the body. Now this is great progress, isn't it! It won't turn rancid in the bottle; thank goodness for that. But once you eat it, it will turn rancid inside your body, because your body cannot break it down with the hydrogen in it.

Now let's discuss protein flesh. With that, you can have green vegetables, the non-starch category. I do not eat protein flesh, but most people who do say that a hot vegetable goes very well with it. So you could have any of those non-starch/green vegetables cooked along with the protein flesh.

The non-starch/green vegetables do not have to be cooked, but here is a "taste" hint. When you eat protein flesh, it eliminates eating fats, acids, or starches. That means if you have a salad you must have it without oil in any form, even in a dressing. Most people do not like salad without some type of dressing. If you are one of those, then this would be a good time to have steamed vegetable. To most people, a piece of chicken sounds better with cooked green vegetables than it does with a salad without dressing.

However, if you are a person who likes raw vegetables without dressing, it would be much better that the vegetables are raw than cooked.

Remember, with food combining, you pick your favorite foods and then just combine them at the right time.

Under the category protein fat, which is cheese, avocado, etc., you can have fat and non-starch/green vegetables *or* you can have acid fruit, if your digestion can handle that. Remember, all nuts should be eaten raw. Once a nut is roasted, it is indigestible.

This is another point I want to stress. Sometimes in stores they will have a machine in which they are making fresh peanut butter. People who buy that think that's real good for them since it's fresh. But you must remember, if those nuts have been roasted, that makes the peanut butter indigestible. So check wherever you're getting your peanut butter to make sure that the nuts are raw, or you cannot digest that peanut butter.

Also, you will notice that peanuts are listed as a starch rather than protein fat. Therefore, many people make a sandwich on bread with peanut butter, but then still develop indigestion. The peanuts must be *raw* to be digestible. Therefore, the peanut butter must be made with *raw peanuts*. Peanuts digest best when sprouted.

Under the protein starch category, because the beans and peas are 50 percent protein and 50 percent starch, they are hard to digest and they should be eaten only with preferably raw, green vegetables. You cannot eat more than one protein type at a meal. Don't mix protein flesh, protein fat and protein starch. You cannot digest them. You have overloaded the digestive tract if you eat more than one at a time. Besides the fact that you overload the enzyme action, if you eat more than one protein at a time, you tend to overeat.

For instance, if you're eating just crab, your taste buds will be tired of crab long before it would if you were having crab and steak, because the two different tastes tend to make you want more and more and more, rather than getting tired of it.

It will take you a while, but once you get in the habit of combining properly, you will be amazed at the food types you can have. Previously, many foods were fattening because you did not digest them. But now, when you start properly combining them, you can have many, many food types you never thought you could have. Remember how many times we ate meat and potatoes, but then started going without the potato because we thought it was the fattening part?

Symptoms of Indigestion

According to the Health Insurance Association there are approximately 100 illnesses caused by indigestion. One of the most common is gas in the abdomen.

One of my clients told me every time she went to a restaurant, no matter what she ate, by the time she left she would have to unzip her slacks. She would leave the restaurant with her slacks slid down from her stomach because she would be so bloated and so uncomfortable from the gas that she couldn't wear them buttoned. Now she's using food combining, and as long as she does it, she never has any problems.

Another common symptom of indigestion is bad breath from the gastro-intestinal fermentation. Bad breath can have other causes, but I have found it to be most commonly from indigestion.

Remember that the digestive tract starts at the mouth and ends at the rectum. If food putrifies *anywhere* between those two areas, it affects the whole body.

Bad breath can be very embarrassing for both the person who has it, and the person being subjected to it. To use breath mints, etc., does not correct the problem, and their odor is only used to temporarily cover up the bad odor.

Bad breath can also be from infected gums and teeth, or from foods left in the mouth between teeth. If you are experiencing this problem, start with extra special care of your teeth. If you don't see an improvement there (and you probably will see some because most people do not clean their mouth as well as they should), concentrate on your digestion more after "specific" foods. If you are combining and digesting properly, you should see the bad breath just fade away.

Many people also have a foul odor in their bowel movements. It's a proven fact after years and years of study, that when people have proper digestion, eat proper foods together and cut out a lot of meat, they do not have foul-smelling bowel movements.

It is "normal" that when meat digests it has toxic residue. Remember you are eating "dead" flesh and "dead" flesh rots. It is a "normal" process.

Therefore, even if you digest your protein flesh meals, you will still have a foul odor in your bowel movements. This odor should help you to realize what is happening inside your body. It is easy to understand that an odor like that could not be creating health inside your body. And believe me, the toxic part that is staying inside is worse than what you are smelling. The longer it is in there, the worse it gets.

Another symptom of indigestion is gas from the rectum. Now, I don't mean air — I mean gas. There is a big difference between *air* and *gas*, and you will be more aware of your digestion problems if you understand that difference. *Air* is what you pass when you're talking to someone, and you go right on smiling and talking. *Gas* is what you pass and you look around for someone else to blame. Sometimes it's so bad and so rotten that the heat from it is just like the heat you feel from the exhaust coming out of the back of a car.

I would like to share my own personal experience with this. By the time I finally discovered that it *does* matter what foods you eat together, it was discovered that my colon was completely prolapsed and lying in my left abdominal area.

For almost a year, I lived on fresh fruit and vegetable juices. Naturally I was full of *gas*. For the whole time I was on the juice program, I passed *huge* pockets of gas, some very hot, some not. However, all had the tremendously foul odor. Now you don't have to tell me about the embarrassment . . . I've had it all.

During that time I did have to make adjustments in my social life. (I almost had to make them in my married life too!) But you aren't going to be that bad. You will pass some foul pockets, but stick with the program because it is merely cleansing your system, and it's better to have that out in the atmosphere than inside your body. Also, the longer you use food combining, and the more junk food you eliminate, the less gas you will have, so be positive that you are just getting better all the time.

Hints on Eating

Do not eat between meals. If you eat, it is your meal. People say, "The only thing I ever eat between meals is fruit." But you see, if you had a meal that takes six hours to digest, and while it's still in the stomach digesting, you send fruit to it, that fruit is delayed from reaching the small intestine and it ferments. So if you eat a meal at noon and about 2 p.m. you decide to have a piece of fruit, that meal is still digesting in the stomach. You send the fruit to it and the fruit's going to ferment and turn to alcohol. If you have fruit, that is your meal. If you have anything, it is your meal. Do not eat between meals. It puts an overload on the digestive system.

If you must eat more often, make sure you eat foods that can be eaten more often. Look at your chart again, and check the digestive times. If you have to eat every two hours, then eat food that digests within two hours. If you have to eat every five hours, find foods that do that.

Keep meals simple. The less combinations of even the proper kind that you serve at a meal, the better.

It always amazes me the length of time that women spend in the kitchen cooking. Some people spend all of Sunday morning in the kitchen cooking because they're preparing this and that, and they have 10 to 12 food combinations, a traditional festive meal. That's way too much work, and it's going to be too much work for your digestive system.

What you want is just the simplest, easiest, best combinations. It saves a lot of stress for the one doing the cooking. It saves a lot of cleanup in the kitchen, and it'll save the whole family digestive problems.

To give you an idea of how much cooking I do, I don't even have a stove in my house. I sold it the week I moved in. I have an electric skillet, a juicer, a blender, a toaster and a crock pot. Stop cooking so much — your whole family's health will improve.

Never have more than one starch at a meal. For instance, if you have baked potato, baked potato is your starch. And you can have two or three potatoes if you'd like. But if you'd rather have homemade sourdough bread without sugar in it, that's your starch. Then you can have other combining foods with it.

But more than one starch is just like more than one protein at a meal. You tend to get in the mood for eating since you don't get tired of the taste and you have two things from which to choose. When you only have one starch, you'll tend to chew it longer, and you won't want to overeat so much.

Note: Because you tend to overeat, it may be that you could eat more than one *if* you measure out your starches before eating so that you are assured of not overeating. Try it if you can't resist sourdough bread with baked potato.

Research shows that at least half of the food people eat leaves the body undigested. Now think about that for a minute. At least half of the food eaten leaves the body undigested. Is it any wonder that people's intestinal tracts are just loaded with poisons, toxins and gas? That very thing is one of the major contributing factors of sickness and death.

Do not eat when you're tired. If you're tired, rest. Take a nap. How many times have you been tired and you thought, "Oh, I guess I'd better eat something."

Sleep is not an alternative for food. Food is not an alternative for sleep. If you're tired, rest.

Don't eat just before beginning hard work. Right after you eat a meal, the body sends an extra supply of blood to the digestive organs. The reason is to come to the aid of the job that needs to be done right away. If you go out and start heavy work, you keep the body from being able to send that extra blood supply to the digestive tract because it needs the blood in your limbs and other parts of your organs for the heavy work.

It takes more energy to digest food than any other function your body performs, so do not think just because digestion is automatic it can be performed under any conditions. It cannot.

Have you ever eaten a meal and been so tired afterwards you can hardly move?

There can be various reasons for that. You may have eaten too much. You may have miscombined. You may need to move around a *little* to get your circulation moving for your digestion to work properly.

It is not a good idea to lie down and take a nap. Many people have died in their sleep doing that because the combustion, caused from indigestion, pressed against the heart so hard that the heart stopped beating.

You also should not take a hearty little walk right after eating. The best thing is just to relax and move around normally, without straining.

Here are some things to notice. Do you notice that same tiredness when you eat only fruit? Do you notice that tiredness when you have starch and vegetables? Do you notice the tiredness when you eat a "major" meal, with protein?

Please try to analyze your own body. It will save

you thousands of dollars in medical bills, and eliminate a lot of illness for you and your family.

Do not eat when you are cold or overheated. For instance, if you come in from doing some yard work and you're very hot and perspiring, take time to cool off before you sit down to eat. And do the same thing if you're very cold. Take time to warm up to a comfortable temperature before you eat.

Don't eat when you're feverish. It shows the body is trying to cleanse. Don't eat when you're in pain; when there is severe inflammation, or when you are not hungry. Have you ever said to yourself, "Well, I'm not really hungry," or "I guess I'd better eat or I'll be hungry later." Well, if you're not hungry, don't eat. You don't need to eat if you're not hungry.

Don't eat when you're worried, anxious, fearful or angry. Eating under all these circumstances favors a reaction of bacterial poisoning of the foods that you're eating. How many times have you been stressed, in a hurry, or upset? You know what you do when you're in that condition? You just shove anything in your mouth. Food will just go in left and right, and you're swallowing it before you even really chew it. These conditions contribute to overeating and poor food combining at the same time. To help you realize the importance of your whole body working together, let me explain something about two nervous systems you have.

Number One: Parasympathetic Nervous System. This system governs the secretion of the digestive juices and the flow of blood into the mesenteric tissue. The mesenteric tissue is a "wall" that lines the backside of the large and small intestines which picks up nutrients (or poisons) and takes them to the liver.

Number Two: Sympathetic Nervous System. This system governs anxiety, worry, and sudden stress. The sympathetic system tends to dominate over the parasympathetic system, and if there is stress during a meal, it will inhibit the body from being able to digest food properly.

If indigestion sets in, you will develop gas, discomfort, etc. Further down in the digestive tract, you develop putrefaction and toxins. Toxins and putrefaction cause fatigue, irritability, headaches and many other unfavorable symptoms. After years of this mistreatment, you can develop colitis, inflammation, constipation, arthritis, high blood pressure, and the list goes on and on. It is best to practice preventative measures such as food combining and nip all these potential problems "in the bud."

Is it starting to make sense to you now? Why every Thanksgiving through Christmas you really feel lousy? Isn't that the truth, though? It's banquet, after banquet, after banquet. It's a proven fact that by New Year's Day people are more ill and more irritable. They're tired, they're run down, they feel awful, and a major cause is an overload of poor food combining along with stressful conditions.

Avoid Salad Bars

Many people trying to lose weight, eat mostly at so-called "salad bars." This could be one of their mistakes.

Have you ever noticed that at home when you grate carrots and other vegetables, they soon turn brown? Yet, in some salad bars, they almost "never" turn brown. It is a common practice to use whiteners, preservatives, etc., on the vegetables so that they will look nice and stay "fresh" for hours (maybe even for days and months). These chemicals are so dangerous if ingested by some persons, that some states have a law that restaurants must post a sign stating they use these chemicals.

Also, at these salad bars, a big percentage of the food is potato salad (poor combining), macaroni salad (poor combining), and fruit (poor if combined with anything). Further, their salad dressings, whether fresh or not, are made with hydrogenated oil, and any flavor of their dressing will contain sugar. So, what is there that you can eat if you want to build your health?

I still say, you are better off to eat just baked potatoes or fresh fruit when eating in restaurants.

I must tell you of an experience I had with salad bars. In my restaurant, I never had a salad bar, even though I wanted one, because the vegetables would turn brown without sulfites and I would not use sulfites. I tried to tell my clients how fattening salad bars are because of the high salt, preservatives, etc. One night I decided to eat at a brand new restaurant. I took my own salad dressing as I knew theirs would be salty. I ate one large serving of salad, with my dressing. The next morning when I arose, I noticed my hands and feet felt swollen and puffy. I stepped on the scale and had gained four pounds from that salad. Oh, were you going to salad bars to lose weight?

I do want you to realize though that you *can* eat in restaurants. But you MUST be aggressive.

One time I took a business man "on the town" for four nights in a row. He refused to believe that he could just order what he wanted, but offered to foot the bill, if I would teach him how.

Our first night we had Chinese food. We had steamed shrimp with pea pods, a large plate each, with no sauce and no salt or M.S.G. (monosodium glutamate — a very fattening appetite stimulant).

Evening two, we ate at a seafood restaurant. We ordered from the regular menu, ate the fish and vegetable, and took the baked potatoes home in our "doggie bag."

Evening three, we ate at an ordinary American restaurant. They had fresh homemade sourdough bread with no sugar, so we had baked potato, bread, steamed zucchini and raw onion and garlic in our baked potato. It was later in the evening, so we had no butter or oils in order to cut down the digestion time. It was all ordered a la carte.

Evening four, we ate Mexican food. Here was the real test. No one knows how to use salt like Mexicans! We had a taco salad with tomatoes, avocado, unprocessed grated cheese and olives, and we did not eat the pastry shell. We paid extra to have plain avocado mashed to use for dressing, instead of the usual guacamole which contains too many "no-no's."

If you're thinking it was expensive to eat this way, you're wrong. Our total bill for the four nights for two was less than he usually spent for one.

The Body Is Not Designed For Complex Meals. It's so interesting the way the whole digestive system works together. The body will get ready for a natural combination.

For instance, if you put starch in your mouth, the stomach and the rest of the body prepare for starch. But if you put protein in your mouth, the rest of the body gets ready for protein. In fact, with protein you don't even have to have the food in your mouth. The body starts getting ready for it as soon as you think about eating a protein.

I mentioned earlier that the body is not designed to handle complex meals. The natural foods designed for the body, meaning fruits and vegetables, are, however, complex foods. They contain fat, protein and starch mixed together. And yet I just told you not to mix these things together. Let's examine what happens when you eat complex foods.

Eating a complex food is your *limit of digestion*. Making a complex meal out of complex foods is *beyond* your limit of digestion. Let's use the potato as an example. The potato is a natural food, and it contains starch and protein and fat. In fact, 1 cup of potato contains 19 grams of starch, 2 grams of protein, and 1 gram of fat. If we can't mix starch and protein, why is it we can eat a potato?

To begin, you put the starch in your mouth and you chew it a long time. While it's in the mouth and you're chewing it, the saliva starts to be produced and it has ptyalin in it. And that ptyalin digests the starch first. The food is then swallowed, goes to the stomach and digestion continues. As the stomach realizes there is protein, the stomach starts producing enzymes to digest the protein part.

Remember, protein digestive enzymes are produced when you *think* about eating it. But with the potato, you didn't think you were eating protein because the potato is a starch food. Therefore, no stomach enzymes for protein were produced until after the starch part was digested. Even then, the protein enzyme amount would be so minute it would not interfere with the starch enzyme. Also, the potato protein is a more compatible protein to starch than, say, an animal protein. Vegetable protein would not require the strong enzyme combinations needed for flesh protein.

It's one thing to eat a complex food and it's quite another thing to eat two or more complex foods of opposite character. Natural combinations require the right amount and type of enzymes which will not cancel each other. The problem develops when foods are radically different.

Develop an Order of Eating. There's an order of food that you might consider also. I always had terrible digestion problems because of extreme health problems. It was finally food combining that brought everything together for me. And even then, I learned an order of foods to eat during each day.

For instance, in the mornings, I only have fruit for breakfast. Sometimes I'm really hungry and I'll eat up to five bananas. Another morning I may not be very hungry and I'll eat one banana, or whatever fruit. But whatever, my only food for breakfast is fruit. It's easy to digest. It helps get a lot of enzymes into the digestive tract. It doesn't make you sluggish because it digests very quickly. Then at noontime when you stop to eat again, if you stop to eat again, breakfast has been digested in plenty of time, and you can have lunch.

Another point is this: When you have not eaten food for over eight to 12 hours, the body automatically goes into a cleansing. Fruits are cleansing foods and since you probably haven't eaten food for over eight hours, when you arise in the morning, it's a perfect time for fruit.

It is very common for people to tell me they are not hungry in the morning, and they just can't eat anything. Stop eating late at night and you will see a big difference. Most people aren't digesting their evening meal. Eat earlier in the evening, and you will have an appetite for a light breakfast.

In the lifestyle that we're in today, most Americans eat breakfast, lunch and dinner. Some eat all three; some just one. But those are the three time periods during the day that they do eat something. I try to eat just two meals a day and have my last meal by 4 p.m. When you do eat, make sure you eat plenty. A common problem in using food combining effectively is that people do not eat enough. Therefore, as a result, while they're waiting during the required time periods in between, they become hungry. Lunch could be more fruit, or a starch with raw or steamed vegetables. Your starch requires five hours to digest so you could still have an evening meal.

Daily Schedule of Meals. We have found the following order of food types to be the best program for a day.

BREAKFAST	FOOD	DIGESTION TIME
Acid or sub-acid fruit	Oranges Peaches	2 Hours
MID-MORNING		
Sweet fruit	Dates	3 Hours
LUNCH		
1. More fruit, or a starch	Steamed rice	5 Hours
2. Starch and green vegetables	Steamed rice Salad Cucumber dressing	5 Hours
DINNER		
1. Starch and green vegetables	Baked Potato, Butter, Green Salad, Oil	10 Hours
2. Protein and green vegetables	Baked chicken, Steamed broccoli	12 Hours
3. Green Salad	with oil dressing	9 Hours

Do not count calories or carbohydrates. When you eat the proper food at the proper time in the right combination, or by itself, you can have a lot more of the foods you like, rather than mixing them with ones you don't like. Be selective. Pick what foods you like best and then find the right time of the day to eat them in the right combinations.

If I have protein, I eat it as my evening meal. The reason is that the body needs a long period for digestion after you eat protein. So does fat. Let's say at noon some day you had a baked potato and you just had to have butter on it. There was your starch, but you put that fat on it and that fat extends your digestion time so long, it should be your last meal for the day. If you eat a food that takes a long time to digest, which would be anything besides fresh fruits and vegetables, then you don't need another meal at the next meal time because you must allow that extra time for the hard-to-digest meal to be digested.

Now what will happen if you don't? Let's say at noontime you ate a salad and you did real well. You had salad oil on it with some lemon juice and all you had was non-starchy vegetables. But remember, you put the oil on it, and that meal is still going to be digesting until 8 p.m. or midnight.

What if you have a big protein dinner at night on top of that? Your previous meal is still in the stomach digesting, and you set up a fermenting action that overloads the digestive tract. Don't rush your meals. Allow plenty of time in between. Don't say, "Well, all I had at noon was one slice of toast with butter on it." It wasn't the volume; it wasn't the amount that mattered. Even though it's a small amount, it is still digesting and the stomach's busy doing that. It does not want another meal thrown on top of it.

Undigested Carbohydrates Ferment. I have a lot of people tell me they never drink alcohol. In fact, maybe it's even against their religious principles.

However, if you eat a carbohydrate, and it ferments in the digestive tract, that fermenting is converted into alcohol. Yes, undigested, fermented carbohydrates in the digestive tract turn into alcohol.

Every once in a while, I hear a case of someone who went to the doctor and the doctor said, "You drink way too much alcohol. You have cirrhosis of the liver. You must stop drinking."

The problem is, the person doesn't even drink; has *never* had a drink. But they don't need to drink. They have their own internal distillery since they are very, very bad about eating sugars, starches, and other foods that stay in the digestive tract too long. When those foods reach the liver, it's the same as if they had gulped down a bottle of alcohol. It's the same poison. If you don't want a liver like an alcoholic, stop eating your carbohydrates at the wrong time.

Complex Foods

I'd like to take you through some foods and discuss their make-up so that you can understand that in their raw form all fruits, vegetables and nuts are complex foods.

In the beginning, Man was commanded to eat fruits and vegetables; things from the trees and grown in the ground. But remember, there is an order in which the body digests everything. When you eat a natural food, the body knows when to digest it and at what stage so that it does not set up a fermenting action.

Also, we don't mix raw foods together just because they're raw. Always refer to your food chart when combining, because raw fruits and raw vegetables cannot be eaten together either.

Let's start with the category of nuts. We're going to use the amount of one cup serving in all these foods that I list.

In one cup of *almonds*, there's 26 grams of protein, 27 grams of carbohydrate and 76 grams of fat.

In 12 *pecans*, there's one gram of protein, one gram of carbohydrate and 10 grams of fat.

You can see that you could easily get unhydrogenated oils from nuts, enough for your daily needs. Because some people say to me, "Well, if I don't have oil, what will I do?" The best source is to eat it in raw nuts or seeds.

In *apples*, there's no protein. There's 15 grams of carbohydrates — and get this — only .04 grams of fat. That's a small amount of fat and since apples are digested in the small intestine, the small amount of fat does not stop the action of the apple being digested quickly. Also, the form of fat in the apple is not the same as in meat, nuts or seeds, and therefore, does not require the 12-hour digestion time.

In *bananas*, there's 1 gram of protein, 23 grams of carbohydrate, .02 grams of fat.

In *blueberries*, there's one gram of protein, 15 grams of carbohydrate and .06 grams of fat. Compare that combination in just blueberries with blueberry pie, and you'll really see what I mean by Man messing up complex foods by making a complex meal.

In *cantaloupe*, there's one gram of protein, 9 grams of carbohydrate and .03 grams of fat.

In *tomato*, there's one gram of protein, one gram carbohydrate and .03 grams of fat. That combination in tomato also helps you realize why it is an acid fruit that can be mixed with protein fat and non-starch/green vegetables. Notice the combination of it compared to the other fruits.

Next, a few vegetables:

In *broccoli*, there's 4 grams of protein, 6 grams of

carbohydrate, .02 grams of fat.

In *potatoes*, there's 2 grams of protein, 19 grams of carbohydrate or starch and .01 gram of fat.

Rice has 2 grams of protein, 23 grams of carbohydrate, .03 grams of fat.

Spinach has 2 grams of protein, 3 grams of carbohydrate and .03 grams of fat.

I want you to notice something about this. When you write all these numbers down, you see one thing in common with all these foods, except nuts: carbohydrates always outweigh the protein.

Think about common, popular diets today. If you want stamina, energy and health, what kind of a diet will most doctors refer to you? High protein. Everybody says, "Oh, I have to have a lot of protein." They have you cut out fats, cut out carbohydrates and eat high protein. And yet, foods designed for the body in the natural form are just the opposite.

You can identify food types for yourself if you find charts that list the starch, protein and fat contents. The thing that determines a starch is if it has more grams of starch than grams of protein or fat. The thing that determines a food to be protein is if it has more grams of protein than grams of starch and fat. The same principle applies to fats.

Making the Change to Proper Food Combining.

If you go back to the original design and you start eating raw fruits and vegetables in the proper food combinations, you will not only have more energy, good health, better digestion, better bowel movements and a better personality, you will also lose weight. You will not only lose weight, you will lose toxic cellulite weight.

For a while, it's going to be necessary that you keep

checking your chart. Build on your favorite combinations.

Eat the foods that you like and remember that fats are fats in food combining. It doesn't matter if they're hydrogenated or unhydrogenated. If you eat a fat in the wrong combination, it will cause putrefaction.

Sugars are sugars in food combining. It does not matter if you use honey or white sugar. If you eat it in the wrong combination, it putrefies.

Starch is starch and it doesn't matter if it's white flour or wheat flour. In the wrong combination, it sets up putrefaction.

Now, please, don't go around saying, "Well, Lee said there's no difference between honey and white sugar." That's not what I said. There is no difference between those in *food combining*.

Find the category of syrups and sugars. You will notice white sugar and honey are under the same category. If you are eat a spoonful of honey that's fine, but you wouldn't eat a spoonful of white sugar. There is a world of difference between white sugar and honey when you're eating them alone.

What I am saying is this. If you eat honey and bread, which is an improper combination, the putrefaction it sets up is no different than if you ate white sugar and bread.

If you eat hydrogenated fat with protein flesh, it will add the same digestion time as unhydrogenated fat and protein flesh. There is no difference in food combining. It's only when you eat them alone, or with the proper food combining that it makes a difference in the quality of the fat, sugar or starch. I hope you understand that, because that's a very deceptive point. Many people have thought as long as they were using fresh ingredients, it would combine. That is not true.

The best thing to realize is, when you're in doubt, just don't mix them.

I remember when I was first doing this. I'd go into a restaurant, having forgotten my chart half the time, and I'd get so confused with the menu. I didn't know what to eat. It took me a while to get used to it, but once you do, you will really enjoy eating.

Here's a good policy to apply when eating in a restaurant. Don't ever read the menu. When the waitress comes over, just tell her what you want.

I owned a restaurant for nine years so I know both sides of the counter. If proprietors are really concerned about their clientele, they want to keep them. Therefore, especially if you frequent the same place often, explain that you cannot eat any of the combinations on the menu and you would like to order *a la carte.*

At this point, the waitress will usually say, "Okay, but I'll have to charge you." *Be extra polite* and say, "Oh, that's okay." (After all, wasn't she going to charge you if you had picked a menu item?)

Then tell the waitress the category you want. For example: You'd like something like rice or baked potato, and you'd like a steamed green vegetable with it. Any nice restaurant will have what you want. Try not to have their breads as your starch because it will probably have sugar in it.

Please do not be afraid to combine just because you travel and eat in restaurants. Just be selective.

Restaurants are famous for salting their foods heavily. Always try to choose foods which do not have a high salt content.

Also, restaurants are famous for using packaged foods, even though they say they are homemade. This applies to their soups, dressings, sauces, and dips. The "homemade" usually means, they open the bag or can, and add the liquid themselves.

We must realize the amount of sulfites used in restaurants is very high. The National Restaurant Association and the Food Allergy Committee released a report showing the following foods in restaurants will contain sulfites:

avocado dip and guacamole
beer
cider
cod (dried)
fruit (cut-up fresh, dried or maraschino-type)
fruit juices
gelatin
purees and fillings
potatoes (cut-up fresh, frozen, dried, or canned)
relishes
salad dressing (dry mix)
salads, (particularly salad bars)
sauces and gravies (canned, dried)
sauerkraut and cole slaw
shellfish (fresh, frozen, canned, dried — including
clams, crab, lobster, scallops, shrimp)
soups
fresh mushrooms
vegetables (cut-up fresh, frozen, canned or dried)
wine vinegar
wine, wine coolers

As a general rule, it is a good idea to eat at least 75 percent of your food in its raw form. Dead food is dead.

Therefore, foods such as meat, eggs, milk products should gradually be eliminated from the diet. Healthy bodies do not crave wrong foods, so as your health improves, you will find you do not desire many foods that you previously did. You will also find in time you will even be able to digest some foods you could not previously digest, because your digestion will be improving. Keep monitoring yourself with the Three-Step Check on page 40.

In Summary.

Let's review the foods you must not eat together, one more time.

- Do not eat acid and starch together.
- Do not eat protein and starch together.
- Do not mix any category of proteins.
- Do not eat acid fruit and protein flesh.
- Do not eat acid fruit and protein starch.
- Do not eat fat and protein flesh.
- Do not eat sugar with anything.
- Never eat fruit with anything, not even other fruits, unless they are in the same category, and even then you must be careful.

Exceptions:

- Some people may be able to eat lemon and tomato with green salad.
- Some people may be able to eat acid fruit with protein fat.

I know you'll benefit from food combining because students and readers call me and say how much better they feel. They're not bloated. They're not as tired. Their health improves. They're losing all that ugly weight. When you get that kind of a reaction within a week, just from changing your eating habits, it shows you how important it is to use Proper Food Combining techniques.

Some Living Testimonies

Here are some typical experiences that students have shared with me.

Man, 52-years-old, lost 26 pounds in three weeks. Blood pressure went from stroke level to normal in five weeks. Discontinued drugs.

Woman, lost two inches from thighs and buttocks in the first week; eliminated craving for sweets, and overcame formerly consistent constipation. During an experiment without Proper Food Combining, she got gas, felt "lousy," and retained water. She returned to the program successfully, and continued to lose weight.

Woman, age 44, eliminated indigestion and lower intestinal distress, improved general feeling of well-being, and within one week got back to size 10 clothes.

Man, 55, snored so loudly that his wife and children had to have television turned up loud to drown out his snoring. Wife decided to leave him. She attended one of my seminars and asked him to try the program. He lost 21 pounds in three weeks, blood pressure dropped from 200/180 to 140/74, and his snoring stopped within two weeks of starting food combining program.

Woman, lost 50 pounds in two months, normalized blood pressure, recovered from burning indigestion, eliminated hemorrhoids and backache, and discontinued all drugs.

Woman, lost 10 pounds in first week (started at 204), lost inches evenly all over, noticed more energy and reduced pain in feet. In first month, lost total of 19 pounds and has begun adapting food combining for her dog.

Man, 58 years old, a pilot who was about to be "grounded" because of his high blood pressure (210/110). Within three weeks it dropped to 124/80. Starting weight was 205 pounds and in 2½ months weighed 168. Lost seven inches in the waist without exercising. He was amazed by results, especially because he was eating *more* food than he previously ate.

For 35 years he had suffered indigestion and acid stomach so acutely that he took Alka-Seltzer *before* he ate, and antacids *after* he ate. From his second day on food combining, he has had indigestion only once, and that was when he ate wrong combinations at a party.

The flight surgeon had given this man one year to live, if he didn't find some way to change his health condition. The flight surgeon didn't change his ways, and he died six months later! Mr. Smith is still alive and flying!

Woman, passed a total of 111 gall stones during the gall bladder cleanse, with no pain, and lost six pounds in the process.

Woman, totally removed hemorrhoids, eliminated asthma, arthritis, varicose veins, and regularized her menstrual cycle. She is constantly told she looks five years younger, and has reduced from size 18 to size 16.

C·H·A·P·T·E·R·4

COMMON QUESTIONS ABOUT FOOD COMBINING

I have collected quite a few questions from readers of the *Proper Food Combining Cookbook* and from people who have attended my seminars. I'd like to share some of the answers with you, because I know many of you will have the same questions.

The first one involves the fact that I have often mentioned that I do not drink milk or eat meat or other dead foods. Many people ask, "If that's wrong, or you feel it's wrong, why does your cookbook have recipes that include milk and meat and cheese and all those?"

First of all, we have to remember, the body does not like fast changes. You cannot instantly decide "Okay, food combining is for me. I'm going to do it —and I'm going to be a vegetarian and, overnight, make that change." The body doesn't like a change like that, either mentally or physically. Many foods that you are eating act as a drug to you. If all of a sudden you were to eliminate them, it can create a withdrawal effect. Sometimes this sets up such a terrible craving for the food that the person may actually return to not only eating it, but actually eating more of it than before.

If you are a person who has already been doing some fasting or you have been on a mono-type diet, you can probably adjust and change right away. But the average person cannot do that.

I have found that many times people grow discouraged if, all of a sudden, they can't do this, they can't do that, and they can't do this. They get so confused by trying to eliminate and change everything all at one time. Don't try to do that. You may be a person who can change fast, but the average person cannot.

Fasting can be excellent for your body if you are healthy enough to do it. Don't start on a health-building program by fasting. Most people are so unhealthy that releasing too much poison, too fast, will literally make them sick. Spend time combining foods properly until you have made it a way of life, and then gradually work into fasting.

In my cassette entitled *Cleanse vs. Surgery* there are some special fasts for different health problems. However, they are fasts of a short duration and you may be able to do one of them.

Those of you who do try fasting, keep in mind that during fasting most of your digestive organs reduce their activities. However, the eliminative organs speed up their activities so there will be more urination, more bowel movements, more toxic gas from the lungs, and more cleansing through the skin. Bathe more often, brush your teeth more often, and keep the bathroom window open.

What I have discovered over time is that clients who have used my food combining gradually, while their body is cleansing and adjusting, find that they no longer care for the foods that were bad for them. That doesn't just include dead protein. It can also apply to chocolate, sugar, and other harmful foods. Once the body is being satisfied with proper foods, and getting the nutrients that it needs, it will not crave the wrong things later.

Don't be in a hurry to adjust to all this overnight. The recipes in the cookbook for protein dishes are there merely to help you combine properly, not to encourage you to eat them. I have also found that some family members who do not want to combine are more cooperative when they can still have foods that they like.

Another question that I am often asked relates to the note at the bottom of each page that reads "for those not combining," followed by a list of foods or recipes that you can have with the recipe at the top of the page. Again, that addition is not to encourage mis-combining. It is to save the combiner a lot of time in the kitchen and to keep the mis-combiner happy at the same time.

When you prepare a meal, fix your Proper Food Combination dish first. If that will not satisfy the other members of your family, add the suggested items at the bottom. That way, you have your dish, and they have what they want, and it makes for a happier family. It will save you a lot of time if one of the dishes that your family eats is the same as one you eat. That's why you want to start with your dish first, and then add the rest of it.

I have had many women tell me that after their family adjusted a little to some of these new menus or dishes, occasionally they would leave out the food that is suggested for the non-combiners. This gradually weans the family members to a proper food combining menu. Use whatever works best for your family.

A lot of people say, "Well, I work all the time and I have my family and everything. I don't have time to work out a diet. I want a diet that will help me lose weight and cleanse and everything."

At the back of my cookbook, there is a two-week menu guide. It gives you the meal, such as breakfast, lunch or dinner. It lists the foods and the page number of the recipe. And it gives you the digestion time. However, if it lists grapefruit for breakfast, it does not have to be grapefruit. It can be another type of fruit. The reason it's listed that way in the diet, is so that you will know to have a fruit for breakfast. It's easy. If you're a busy woman who has a family and works outside the home, having fruit in the morning makes it much easier for you than to have to fix some kind of a breakfast that requires cooking.

Note: If you should start cleansing rapidly after beginning to use food combining, you may develop one or two acid sores in the mouth. If so, use fruits which do not cleanse as fast, such as "sub-acid" or "sweet" fruits.

Any of the recipes listed for lunch or for dinner can be exchanged with the same food type. For instance, if for lunch it mentions to have a baked potato, you could substitute rice. It must only be prepared in the same manner. If you were going to have rice instead of a baked potato at noontime, you would not put butter on the rice if you would not put butter on the baked potato. The two-week menu guide can be repeated every two weeks. You will not be deficient in nutrients, and it will make it very easy for you if you are a person who has a problem arranging menus.

Another common question relates to taking digestive enzymes with meals. If you do, they should be considered only a temporary aid until you see an improvement in your digestive tract. Also, be sure to take enzymes compatible with what you are eating. For instance, hydrochloric/pepsin tablets should be taken with protein. Pancreatic could be taken with starch, protein, or non-starch vegetables.

However, be aware that you are trying to improve your own digestive ability, so check yourself occasionally to see if you can digest the food without added enzymes. If you are using Lee Tee, you will see a vast improvement in your digestion, and Trim #1 and #2 will help your body correct digestion and elimination problems.

You should not need enzymes with raw fruit, as they have their own, and digest very easily.

Another question I hear often is, "How do I get my children to food combine?" I realize this is one of the major problems of diet in this country. Children are so geared to fast food.

I recently read a report (*Newsweek,* January 20, 1986) that showed that the number of home-cooked meals prepared by working mothers and mothers who do not work outside the home is almost identical. That means that all children are getting a large percentage of fast food meals.

One thing I have found with children is that we cannot expect them to use food combining if we don't. Probably the most important thing you will do is set an example for the children in the way that you eat.

Another consideration is: adults don't like to be forced into anything, and neither do children. However, children are very adaptable, and I find that they will change to food combining much faster than adults, if it is explained to them what is happening; what is wrong with eating poorly combined foods; what happens when it putrefies; how ill they will become from it; how they can handle society. These are all things affected by diet.

On the other hand, you don't want to be too easy with children. If there were a dish of strychnine in the middle of the table and your child wanted it, would you say, "Well, okay. You can have a little bit of it today, but you can't have any tomorrow?" No.

So you must realize that you are the parent. You're

the one who should make the decision on what they eat. Although you don't want to force them to do it all the time, it will help if you use a firm hand in helping them to realize that you are changing because you love them.

Another thing you might try is to stop rewarding your children with poison. So many times I hear this, "Well, Johnny was just real good this week. He did all his chores and he didn't give me any trouble at all, so Sunday we all went down to the ice cream shop and had a hot fudge sundae." You reward the child with poison.

There are many other things besides food which could be used as a reward, and there are many foods besides poisons. So just think about some of these things that have become habits with all of us, and it probably will make a big improvement with your children.

Another question people often ask is, "How about protein? In food combining, how can I get more protein?"

Maybe we should stop and realize that actually, this country is getting far too much protein. Let me give you an example of that. At birth, a mother's nursing milk is approximately 2.3% protein. That's a very small amount of protein. Six months later, when the child has either doubled or tripled in size, the protein in the nursing milk has been reduced by almost half. We have all been told that we need protein so the body can make cells. Yet during the life of a human, when would a person manufacture more body cells than during this six months of life? There would be no time that could match it, and yet during that time, nature reduces the protein by almost half in the baby's food.

Another interesting point is that mother's nursing milk, at the end of that six months, and green vegetation have almost the same exact ratio of protein.

Another common question relates to hypoglycemia. Hypoglycemics have been told they must eat often, and they must eat a lot of protein. The most up-to-date, current facts show that the hypoglycemic feels better because protein acts as a drug to this condition. (*The Golden Seven Plus One,* by C. Samuel West.) It is not because the protein corrects the hypoglycemia problem. It is because it acts as a drug. Anyone who's been on a drug knows that as soon as you take that drug, you feel better. And that's what's happening with the hypoglycemic. It takes a little adjusting, but since the hypoglycemic does have to eat more often, find the categories of food and pick the ones that you can eat as often as you need. To begin with, you're going to find that that's mainly fruit, and yet a hypoglycemic is told not to eat fruit.

I have found that the change is easier if you use my Trim herbal cleansing program and Lee Tee, along with food combining. The cleansing action of the herbs helps the body rejuvenate itself faster. I recently talked to a woman who had been suffering from severe hypoglycemia. The first day that she was on my program, she was "shakey" all day long. (She did not use Lee Tee.) She kept eating fruit every two hours. We decided we would work it through together for a 24-hour period, and the next day she was over the shakey phase. She rested a lot the first day. Make sure you do not overextend yourself during the first day you try to adjust to Proper Food Combining. Go slow!

Because I am a firm believer that vitamins should be supplements to a proper food program, rather than a food program being a supplement to vitamins, I have searched for *years* for a healthy "fast" food that people of all ages could easily consume, and that requires almost no preparation. I have found it!

The food is called Lee Tee (cereal grass juice). Heating and/or freezing has been proven to change the molecular structure of food enough (even if it doesn't destroy the vitamins, it changes the structure) that the body cannot get the value from them that it should. The process used to dry the barley leaf and wheat grass juice is a patented quick dry method, at room temperature, which dries the juice in two to three seconds, without heat. Therefore, the enzymes are not destroyed, and the molecular structure is not affected.

Earlier in this book, I discussed the importance of enzymes. They perform thousands of functions in the body. Heating higher than body temperature destroys the enzymes and you have lost one of the most important components in your food. Lee Tee contains over 100 enzymes which makes it almost "predigested" before you swallow it.

Naturally, the taste of anything is going to affect how much of it we want to swallow too, isn't it? Because the grass is harvested at just the right time, there is no bitter taste. To me, it tastes like parsley. Others say it tastes like fresh spinach.

A food like this is so *necessary*. Grocery stores only allow about 10 percent of their space for fresh foods. Health food stores usually sell *only* packaged foods and supplements.

It is true you can purchase many things that do not have preservatives or food coloring, or white sugar, or white flour, etc. However, that is all foods that have the *do nots*. We need a food that has the *do's*. Because many people are familiar with certain foods as a good source for certain vitamins or minerals, I want to compare some with Lee Tee:

Lee Tee has:

- twice as much *protein* as wheat germ.

- 13 times more *potassium* than wheat germ.
- 13 times more *calcium* than wheat germ.
- 10 times more *vitamin E* than rice.
- three times more *protein* than rice.
- 37 times more *potassium* than rice.
- 59 times more *calcium* than rice.
- seven times more *vitamin C* than oranges.
- five times more *iron* than spinach.
- 10 times more *calcium* than milk.
- five times more *copper* than spinach.
- 433 times more *carotene* than milk.
- 25 times more *potassium* than bananas.
- twice as much *magnesium* as wheat flour.

Lee Tee also helps to balance the acid/alkaline ratio.

For those who have digestive problems, Lee Tee is absorbed from the mucous membrane in the mouth, and through the intestinal wall. It is digested almost as soon as you swallow it. With food combining, it may be taken just a few minutes prior to a fruit meal, even though it is from vegetation.

Now we know how ancient Man was so healthy on a food combining program without all the big meals of modern day diets. He could eat very often if he chose, because with live enzymes filling his foods, they digested much, much faster.

I have found that for the person with a blood sugar problem, Lee Tee and Trim #1 and #2 together have done wonders to help the body stabilize itself.

True hunger comes from a drop in blood sugar, so if you or your children are hungry in between meals, feel free to drink a glass of Lee Tee. It will give you energy, because, in addition to enzymes, it also contains 16 vitamins, 23 minerals, 18 amino acids (in the easily assimilable form, not dead food), chlorophyl, and, of course, hundreds of enzymes. Lee Tee is truly a total food.

Modern day food is deficient in most of the ingredients of which Lee Tee has high ratios. You only need to try it for a few weeks, and your energy and stamina increase alone will show you the power contained in this powder.

Many people question if they can still take vitamins when using food combining. Yes, you may. But you will probably find that you do not need as many as they did before. Some people completely stop their vitamins the first month of food combining just so that they can see if they do still need all of the kinds they are taking. As you become more aware of feelings in your body, you will be able to regulate your vitamins better. For instance: if you previously developed a cold quite often, you probably took vitamin C to help correct that problem. If on food combining you find that you seldom have a cold, you would determine you probably don't need as much vitamin C.

I also receive many questions regarding different health problems. I am not doing private consultation. The main purpose of this book is to help people learn about their bodies so they can understand what symptoms to watch for and what things that they can do to improve their health.

You need to be able to know, when you have a symptom of something, whether it is from indigestion, or from something else. Many people are spending thousands of dollars going to doctors and other treatments that do them absolutely no good, because the root of their problem is the putrefaction caused by improper food combining. Believe me, you are not going to starve if you cut down on the number of foods you eat at one time. You're not going to starve if you have starch at one meal, and have to wait for the next meal to eat protein.

Some Final Thoughts:

If you are really concerned about your health and the care of your family's health, you must start making some changes. The statistics show that 90 percent of adult Americans are overweight. If you go to a recreation area where there are a lot of children, watch the children and you will see they are not far behind. Many children today are overweight.

We see all kinds of problems with the younger generation due to not being able to adapt socially. Their social behavior is so bad, many people today are having their children committed for psychiatric care. and in thousands of cases, they are made to realize that the root of the problem mentally and physically with the children was not society, but was merely the child's diet.

We repeatedly see transformations in people's lives when the only change that they have made in their routine is their diet. It makes us stop and think. We must do something about it. Over $60 billion dollars a year is spent on medical care in this country. *$60 billion.* And the government's report in 1977 showed that six out of 10 causes of death in this country are caused by improper diet. A large percentage of that $60 billion dollars could be saved, if people just applied their brain to learn how to feed their body.

With food combining, the only thing you're doing is eating food according to the original design of the body. We cannot change the design of the body. That has been tried for years. Parts have been cut out, new organs put in that are metal or from someone else's body. That has not improved the design of the body.

But Proper Food Combining, to insure your body gets nourishment, can help improve your health. And it's not expensive. You buy groceries anyway. Why not buy and eat together the ones that will benefit your body instead of poisoning it?

Actually, most people tell me they save money on their grocery bill once they start food combining. That's because eventually, as you eliminate garbage like prepackaged foods, which are the most expensive in the grocery store, you replace them with fresh fruits and vegetables, which are much less expensive.

Finally, many people are on drugs and they want to know if they should continue taking those drugs while they are using food combining.

As I mentioned before, the body does not like fast changes. If you are going to eventually eliminate drugs, do it gradually.

Drugs should never be stopped *immediately* if a person has been on them for a period of time. The body hates quick changes and they should be gradually tapered out of your system if you are going to try to discontinue use of them. Some doctors have said that good nutrition makes the drugs work even better.

Check with your doctor first before you start eliminating the medication. If, for example, your blood pressure has reached normal, ask why you should stay on the drug.

I have been experimenting with Proper Food Combining for many years. I always felt better while eating that way, but the popular diet of the world would often attract me again, and I would fall back to old habits, until I felt bad and gained a lot of weight. Those two realizations would shake me back into a Proper Food Combining diet. After years of research and experiences from hundreds of my clients, I became very serious about helping myself, my family, and thousands of others to understand food combining. With this knowledge, I have adopted this as a way of life, and have benefited from it in every way.

Good health to you and your family. Hopefully, you will also become "Living Testimony" that Proper Food Combining WORKS!

For a free information packet regarding the products mentioned in this book, please write or call for the name and address of your closest distributor.

Lee DuBelle
P.O. Box 35860
Phoenix, Arizona 85069
(602) 863-2715

INDEX

Other Information and Resources from Lee DuBelle

Proper Food Combining Cookbook: This cookbook contains many mouth-watering recipes, applying the basic principles of Proper Food Combining. It also includes a two-week menu guide. Spiral bound for easy use in the kitchen. (216 pages).

Internal Cleansing is an Old Movement: Learn ways to eliminate the toxins that have built up in your body over the years in order to restore the functions of your vital organs. This soft-bound book provides step-by-step instructions, as well as case studies (120 pages).

Combinando Correctamente Los Alimentos Funciona: This is the spanish translation of *Proper Food Combining Works*, produced especially for those whose first language is Spanish. (120 pages).

Proper Food Combining Charts: These laminated charts are produced in color and are invaluable for anyone interested in Proper Food Combining. They are available in two sizes: *Medium* (12"x9" folds to 6"x9") – can be slipped into book or tape album; and *Small* (6"x3.5" folds to 2"x3.5") – business card size, so that you can carry your food combining information with you.

Lee DuBelle's Cassette Tape Albums:

Series One: *"Proper Food Combining/Gaining Health, Losing Weight"* introduces Proper Food Combining techniques in a clear, easy-to-understand way, and discusses obesity as a disease that must be treated to produce increased health and well-being.

Series Two: *"Cleansing vs. Surgery"* demonstrates that cleansing organs is far superior to removal of them. It teaches the function and purpose of each organ and its importance to the body's operation and immune system.

Series Three: *"The Pre-Menstrual Syndrome/ Cellulite Connection"* discusses the interrelationships of PMS, menopause, hormones, cellulite, as well as how the endocrine system works.

Series Four: *"AIDS, Yeast Infection and Other Immune Diseases"* explores how the immune system works and what you can do to help the body build, or rebuild, its own immune system to prevent illness.

Lee DuBelle's Exercise Video: This 30-minute video teaches Lee's own personal exercise program utilizing gentle physical exercises and mild aerobics in conjunction with a slant board and mini-trampoline. the workout program is designed to tone the body's muscles and tissues to improve the operation of prolapsed organs and the immune system.

Proper Food Combining Video: This exciting and informative 70-minute video explains in detail how Proper Food Combining works. Learn to use the food combining chart, to recognize foods of each category, and to prepare easy recipes.

Colon Therapy Video: Lee made this 30-minute video with muscle therapist, Ron Geschwentner, to help you learn self colon therapy techniques. A valuable resource for anyone suffering from constipation, diarrhea, gas, diverticulitis or prolapsed colon, it includes laminated instructions for internal washing (enema).

LEEWEIGH DIET Cassette Tape Album: Finally a diet that builds health while you reach your desired weight and size. A diet according to the "body's design". A diet for *permanent* weight control. You've tried all the ones that don't work – now try the one that DOES!! Complete hour-by-hour menu and instructions.

ORDER FORM
Lee DuBelle
P.O. Box 35860
Phoenix, Arizona 85069

QUAN.	DESCRIPTION	PRICE	TOTAL
_____	copies of *Proper Food Combining Works: Living Testimony*	$9.00	______
_____	copies of *Combinando Correctamente Los Alimentos Fonciona*	$9.00	______
_____	copies of *Proper Food Combining Cookbook*	$15.00	______
_____	copies of *Internal Cleansing is an Old Movement*	$9.00	______
_____	copies of laminated food combining chart		
	Medium (12 x 9 folds to 6 x 9)	$5.00	______
	Small (folds to business card size)	$3.00	______
_____	copies of audio cassette album #1 (4 tapes/6 hours) "Proper Food Combining/Gaining Health, Losing Weight"	$40.00	______
_____	copies of audio cassette album #2 (4 tapes/4 hours) "Cleansing vs. Surgery"	$40.00	______
_____	copies of audio cassette album #3 (4 tapes/6 hours) "The Pre-Menstrual Syndrome/Cellulite Connection"	$40.00	______
_____	copies of audio cassette album #4 (4 tapes/3 hours) "AIDS, Yeast Infection, and Other Immune Diseases"	$40.00	______

Total this page ________

(continues on next page)

_____ copies of audio cassette album (2 tapes/90 minutes) "LEEWEIGH Diet" $25.00 _____

_____ copies of Exercise Video. (30 minutes).VHS $30.00 _____

_____ copies of Proper Food Combining Video. (70 minutes).VHS $50.00 _____

_____ copies of Colon Therapy Video. (30 minutes). VHS $30.00 _____

Total from previous page __________

Sub-Total __________

Add shipping and handling – $4.00 first item, 75¢ each additional item

Canada: add $6.50 to above total _______

Arizonans, please add 6.7% sales tax _______

US FUNDS ONLY

TOTAL DUE _______

Method of Payment: Check ❑ Money Order ❑

Name ______________________________

Address ______________________________

City ______________ State ________ Zip __________

Phone Number ______________________________

❑ Please send more order forms.

❑ Please send free information packet.

ORDER FORM
Lee DuBelle
P.O. Box 35860
Phoenix, Arizona 85069

QUAN.	DESCRIPTION	PRICE	TOTAL
_____	copies of *Proper Food Combining Works: Living Testimony*	$9.00	______
_____	copies of *Combinando Correctamente Los Alimentos Fonciona*	$9.00	______
_____	copies of *Proper Food Combining Cookbook*	$15.00	______
_____	copies of *Internal Cleansing is an Old Movement*	$9.00	______
_____	copies of laminated food combining chart		
	Medium (12 x 9 folds to 6 x 9)	$5.00	______
	Small (folds to business card size)	$3.00	______
_____	copies of audio cassette album #1 (4 tapes/6 hours) "Proper Food Combining/Gaining Health, Losing Weight"	$40.00	______
_____	copies of audio cassette album #2 (4 tapes/4 hours) "Cleansing vs. Surgery"	$40.00	______
_____	copies of audio cassette album #3 (4 tapes/6 hours) "The Pre-Menstrual Syndrome/Cellulite Connection"	$40.00	______
_____	copies of audio cassette album #4 (4 tapes/3 hours) "AIDS, Yeast Infection, and Other Immune Diseases"	$40.00	______

Total this page ________

(continues on next page)

_____	copies of audio cassette album (2 tapes/90 minutes) "LEEWEIGH Diet"	$25.00	_____
_____	copies of Exercise Video. (30 minutes).VHS	$30.00	_____
_____	copies of Proper Food Combining Video. (70 minutes).VHS	$50.00	_____
_____	copies of Colon Therapy Video. (30 minutes). VHS	$30.00	_____

Total from previous page ____________

Sub-Total ____________

Add shipping and handling – $4.00 first item, 75¢ each additional item

Canada: add $6.50 to above total ________

Arizonans, please add 6.7% sales tax ________

US FUNDS ONLY

TOTAL DUE ________

Method of Payment: Check ❑ Money Order ❑

Name ____

Address______________________________________

City ____________________ State ________ Zip ____________

Phone Number________________________________

❑ Please send more order forms.

❑ Please send free information packet.